Lecture Notes in Mathematics

Edited by A. Dold and B. Eckmann

562

Roe W. Goodman

Nilpotent Lie Groups:
Structure and Applications to Analysis

Springer-Verlag
Berlin · Heidelberg · New York 1976

Author
Roe William Goodman
Department of Mathematics
Rutgers
The State University
New Brunswick, N. J. 08903/USA

Library of Congress Cataloging in Publication Data

Goodman, Roe.
Nilpotent lie groups.

(Lecture notes in mathematics ; 562)
Bibliography: p.
Includes index.
1. Lie groups, Nilpotent. 2. Representations of groups. 3. Differential equations, Hypoelliptic. I. Title. II. Series: Lecture notes in mathematics (Berlin) ; 562.
QA3.L28 no. 562 [QA387] 512'.55 76-30271

AMS Subject Classifications (1970): 44A25, 17B30, 22E25, 22E30, 22E45, 35H05, 32M15

ISBN 3-540-08055-4 Springer-Verlag Berlin · Heidelberg · New York
ISBN 0-387-08055-4 Springer-Verlag New York · Heidelberg · Berlin

Printed in Germany
Printing and binding: Beltz Offsetdruck, Hemsbach/Bergstr.

Table of Contents

Preface

These notes are based on lectures given by the author during the Winter semester 1975/76 at the University of Bielefeld. The goal of the lectures was to present some of the recent uses of nilpotent Lie groups in the representation theory of semi-simple Lie groups, complex analysis, and partial differential equations. A complementary objective was to describe certain structural aspects of simply-connected nilpotent Lie groups from a "global" point of view (as opposed to the convenient but often unenlightening induction-on-dimension treatment).

The unifying algebraic theme running through the notes is the use of filtrations; indeed, nilpotent Lie algebras are characterized by the property of admitting a positive, decreasing filtration. The basic analytic tool is a homogeneous norm, which replaces the usual Euclidean norm and gives a "non-isotropic" measurement of distances. One obtains a filtration on the algebra of germs of C^{∞} functions at a point by measuring the order of vanishing in terms of the homogeneous norm. This in turn induces a filtration on the Lie algebra of vector fields and on the associative algebra of differential operators. To construct (approximate) inverses for certain differential operators, one uses integral operators whose order of singularity along the diagonal is measured via the homogeneous norm.

A recurring aspect of our constructions is the approximation of one algebraic structure by a simpler structure; e.g. a filtered Lie algebra is approximated by the associated graded Lie algebra; a group germ generated by a Lie algebra of vector fields is approximated by the Lie group generated by a "partially isomorphic" nilpotent Lie algebra. The "order of approximation" is sufficiently good that results in analysis on the simpler structure can be transfered to corresponding results on the original structure; e.g. convolution operators on a filtered nilpotent group which are "singular at infinity" have the same L^p-boundedness properties as operators on the corresponding graded group.

The notes are organized as follows: Chapter I studies nilpotent Lie algebras and groups viewed as locally nilpotent derivations and locally unipotent automorphisms of filtered polynomial rings. Comparisons, both algebraic and analytic, are made between various nilpotent group structures. These constructions are continued in the Appendix, in the context of groups of birational transformations.

In Chapter II we explore the possibility of approximating a finitely-generated (infinite-dimensional) Lie algebra of vector fields by a (finite-dimensional) graded nilpotent Lie algebra. This leads to the notion of "partial homomorphism" of graded Lie algebras, and the problem of "lifting" a partial homomorphism. The prototype for this situation is the case of a homogeneous space for a group, where the "lifting" is obtained by identifying functions on the homogeneous space with functions on the group which are left-invariant under the stability subgroup of a fixed point. The main result of this chapter is that a similar construction can be carried out relative to a partial homomorphism which is "infinitesimally transitive". In concrete terms, this means that if one wants to study a set of vector fields on a manifold which have the property that their iterated commutators span the tangent space at each point of the manifold, then for local questions it suffices to consider the case in which the manifold is a nilpotent Lie group, and the vector fields are "approximately" left invariant. We describe how vector fields of this type arise in connection with real submanifolds of complex manifolds.

Chapter III is devoted to constructing a theory of "singular integral operators" which is sufficiently general to include the "approximate convolution" operators associated with the "approximately-invariant" vector fields of the previous chapter. We prove the boundedness of these operators on L_p, $1 < p < \infty$, and study the interaction between the integral operators and the vector fields.

Chapter IV contains the applications to the representation theory of real-rank one semi-simple Lie groups, harmonic analysis on generalized upper half-planes, and regularity properties of hypo-elliptic differential operators associated with transitive Lie algebras of vector fields.

It is obvious from the comments and references at the end of each chapter that a large part of these notes is based on the work of E.M. Stein and his collaborators. We have tried to make the notes as self-contained as possible, by giving complete proofs of almost all results. In the comments and references an attempt has been made to assign credit to the various results and proofs; we apologize in advance to authors whose work we have overlooked or miscredited.

The author would like to thank Prof. Horst Leptin for the invitation to visit the University of Bielefeld and for his gracious hospitality. To Prof. Leptin and his collegues in the mathematics department go many thanks for making the visit stimulating and enjoyable. Dr. Peter Müller-Römer provided invaluable assistance in organizing the manuscript, and the splendid job of typing was done by Frau Mönkemöller.

Much of the writing of the manuscript was done while the author was visiting at the Institut des Hautes Etudes, Bures-sur-Yvette, France. The author thanks the Institut for its hospitality, and Rutgers University for providing financial support through its sabbatical leave program. Several sections of the manuscript are based on reseach supported by grants to the author from the National Science Foundation.

Bures-sur-Yvette

May, 1976.

Notational Conventions An attempt has been made to keep the notations consistent from one chapter to the next; the reader's indulgence is requested for the lapses from uniformity which crop up. Each section is numbered serially, and generally contains at most one Lemma, Proposition, or Theorem. In subsequent sections the Theorem of § 3.3 of Chapter I, say, is called Theorem I.3.3. As usual, $\mathbb{N}$, $\mathbb{Z}$, $\mathbb{R}$, $\mathbb{C}$ denote the non-negative integers, integers, real numbers, and complex numbers, respectively.

Structure of nilpotent Lie algebras and Lie groups

§ 1. Derivations and Automorphisms of filtered polynomial rings

In this section we will construct a class of nilpotent Lie algebras and Lie groups. Every nilpotent Lie algebra and simply-connected Lie group can be embedded in this class. These "model" algebras and groups can be presented either as derivations and automorphisms of a polynomial ring, or dually as vector fields and non-linear transformations on a vector space. This duality will be used in passing from algebraic to analytic properties of the embedded algebras and groups.

1.1 Dilations and Gradations. Let V be a finite-dimensional real vector space, and let $P = P(V)$ be the algebra of real-valued polynomial functions on V. We take as our basic datum a direct-sum decomposition

$$V = \sum_{n=1}^{r} \oplus V_n .$$

(We allow the possibility that $V_n = 0$ for some n.) In terms of this decomposition, we define a one-parameter group of dilations $\{\delta_t : t > o\}$ on V by setting

$$\delta_t(\Sigma\, x_n) = \Sigma\, t^n x_n \qquad (x_n \in V_n) .$$

We then define the space H_n of homogeneous polynomials of weight n by

$$H_n = \{f \in P : f \cdot \delta_t = t^n f\} .$$

Fix a basis $\{x_i : 1 \leq i \leq d\}$ for V such that $x_i \in V_{n_i}$, and let $\{\xi_i\}$ be the dual basis for V^*. If $\alpha \in \mathbb{N}^d$ is a multi-exponent, write

$$\xi^\alpha = \xi_1^{\alpha_1} \cdots \xi_d^{\alpha_d} , \quad w(\alpha) = \Sigma\, n_i \alpha_i .$$

Since $\xi_i \in H_{n_i}$, we have $\xi^\alpha \in H_{w(\alpha)}$. Thus

$$P = \sum_{n \geq o} \oplus H_n$$

Obviously $H_m H_n \subseteq H_{m+n}$, so the spaces H_m give a gradation on the algebra P. Note that when $1 < n \leq r$, there are both linear and non-linear polynomials in H_n.

We will also need the filtration $\{P_n\}$ on P obtained by setting

$$P_n = \sum_{k \leq n} H_k .$$

Evidently

$$\mathbb{R} = P_0 \subseteq P_1 \subseteq P_2 \subseteq \cdots, \qquad \bigcup_{n \geq 0} P_n = P$$

$$P_m \cdot P_n \subseteq P_{m+n} .$$

We shall call P_n the space of polynomials of weight $\leq n$.

Remark. Let $F_n = \sum_{k \geq n} V_k$. The subspaces $\{F_n\}$ are then a decreasing filtration of the vector space V. We claim that the filtration $\{P_n\}$ is uniquely determined by $\{F_n\}$. Indeed, if we define

$$F_{*n} = F_{n+1}^{\perp} \subset V^* ,$$

then

$$F_{*1} \subseteq F_{*2} \subseteq \cdots \subseteq F_{*r} = V^* ,$$

and

$$P_n = \sum F_{*i_1} \cdots F_{*i_k} ,$$

with the sum taken over all indices $i_1 + \cdots + i_k \leq n$.

1.2 Homogeneous norms The passage from analysis on $\mathbb{R}^n$ to analysis on nilpotent Lie groups begins by using the dilation group δ_t in place of scalar multiplication by t. The next step is to replace the Euclidean norms by the following functions:

Definition A dilation-homogeneous norm on V is a continuous function $x \longmapsto |x|$ such that

(i) $|x| \geq 0$

(ii) $|x| = 0 \iff x = 0$

(iii) $|\delta_t x| = t\,|x|$ for all $t > 0$.

The norm is smooth if it is C^∞ away from 0, and it is symmetric if $|x| = |-x|$.

Example Let $0 < p < \infty$, and let $\{\xi_i\}$ be as in 1.1. Then

$$|x|_p = (\Sigma|\xi_i(x)|^{p/n_i})^{1/p}$$

$$|x|_\infty = \max_i \; |\xi_i(x)|^{1/n_i}$$

are symmetric homogeneous norms. If p is divisible by $2r!$, then $(|x|_p)^p$ is a polynomial, so that $|x|_p$ is real-analytic away from zero in this case.

In this chapter we shall be using homogeneous norms to measure the rates of growth or vanishing of functions. The following simple facts will be needed:

I. If $|x|$ is any homogeneous norm, then there exists $C > 0$ such that

$$C^{-1}\,|x|_\infty \leq |x| \leq C\,|x|_\infty .$$

Hence all homogeneous norms are equivalent. (The set $|x|_\infty = 1$ is compact and hence $|x|$ is bounded above and below on this set.)

II. If f is a polynomial, then $f \in H_n$ iff $|f(x)| \leq C'|x|^n$. (Indeed, if $f \in H_n$, then

$$|f(x)| \leq (\max_{|u|=1}|f(u)|)\,|x|^n .$$

Conversely, write $f = \Sigma f_k$, with $f_k \in H_k$. Then $f \circ \delta_t = \Sigma t^k f_k$. But by assumption $f \circ \delta_t = O(t^n)$, so letting $t \to \infty$ we get $f_k = 0$ for $k > n$, and letting $t \to 0$ we get $f_k = 0$ for $k < n$.)

III. If φ is a C^∞ function defined on a neighborhood of 0 in V, then for every $n \geq 0$ there is a unique polynomial $p_n \in \mathcal{P}_n$ such that

$$|\varphi(x) - p_n(x)| \leq C\,|x|^{n+1}$$

for x near 0. (The polynomial p_n is just the sum of the terms of weight $\leq n$ in the Taylor series of φ.)

1.3 Vector fields with polynomial coefficients Let $\mathrm{Der}(\mathcal{P})$ be the Lie algebra of derivations of the algebra $\mathcal{P}$. Since $\mathcal{P}$ is generated by 1 and the linear functions, any derivation T is determined by its restriction to V^*. If we set $p_i = T(\xi_i)$, then

$$T = \Sigma\, p_i\, \partial/\partial\xi_i \, .$$

Hence $\mathrm{Der}(\mathcal{P})$ may be identified with the Lie algebra of vector fields with polynomial coefficients on V.

Using the filtration $\{\mathcal{P}_n\}$, we define the following subalgebra of $\mathrm{Der}(\mathcal{P})$:

$$\underline{n} = \{X \in \mathrm{Der}(\mathcal{P}) : X\mathcal{P}_n \subseteq \mathcal{P}_{n-1}, \forall n\} \, .$$

To determine the structure of $\underline{n}$, we write D_x for directional derivative in the direction $x \in V$:

$$D_x f(y) = \frac{d}{dt}\Big|_{t=0} f(y+tx) \, .$$

Denote by $\mathcal{D}_n = \{D_x : x \in V_n\}$. Then it is immediate that

$$\mathcal{D}_n\, H_k \subseteq H_{k-n} \, .$$

From this it is clear that

$$\underline{n} = \sum_{0 \leq k < n} \mathcal{H}_k \mathcal{D}_n .$$

Set

$$\underline{n}_k = \sum_{m=k}^{r} \mathcal{H}_{m-k} \mathcal{D}_m .$$

Then one easily verifies that

$$\underline{n}_j \mathcal{H}_k \subseteq \mathcal{H}_{k-j}$$

$$[\underline{n}_j, \underline{n}_k] \subseteq \underline{n}_{j+k} .$$

Thus we have a <u>gradation</u> of $\underline{n}$:

$$\underline{n} = \sum_{k=1}^{r} \underline{n}_k ,$$

and $\underline{n}$ is obviously a finite-dimensional nilpotent Lie algebra.

Suppose now that Ω is an open neighborhood of 0 in V. Let $C = C^{\infty}_{\mathbb{R}}(\Omega)$ be the algebra of real-valued smooth functions on Ω. We will measure the rate of vanishing of functions at 0 in terms of the homogeneous norm $|x|$. If $m \geq 0$, set

$$C_m = \{f \in C : f(x) = O(|x|^m) \text{ near } 0\} .$$

(For $m < 0$, set $C_m = C_0$). This defines a <u>filtration</u> on C:

$$C = C_0 \supset C_1 \supset \cdots ,$$

$$C_m \cdot C_n \subseteq C_{m+n} .$$

It follows by the graded from of Taylor's series that if $n \geq 0$, then

$$C_n = \mathcal{H}_n \oplus C_{n+1}$$

and

$$\mathcal{D}_k C_n \subseteq C_{n-k} .$$

Let $L = L(\Omega)$ be the Lie algebra of C^∞ vector fields on Ω. If $T \in L$, we shall say that T __is of order__ $\leq k$ __at__ 0 if

$$T\, C_n \subseteq C_{n-k} \quad \text{for all } n .$$

Set

$$L_k = \{T \in L : T \text{ is of order} \leq k \text{ at } 0\} .$$

For example, $D_k \subset L_k$. This definition formalizes the notion that if a function vanishes to order n at 0, then its derivatives of "weight" k vanish to order $n-k$ at 0.

__Proposition__ The following properties hold:

(a) $L = L_r \supseteq L_{r-1} \supseteq \cdots$

(b) $[L_j , L_k] \subseteq L_{j+k}$

(c) For every $k > 0$, $\underline{n}_k \subset L_k$ and

$$L_k = \underline{n}_k \oplus L_{k-1} .$$

__Proof__ We can write $L = C_0 D_1 \oplus \cdots \oplus C_0 D_r$. Hence $L\, C_n \subseteq C_{n-r}$. This gives (a). Part (b) follows from the definition of L_k. For part (c), we have

$$H_m D_k C_n \subseteq H_m C_{n-k}$$

$$\subseteq C_{m+n-k}$$

Taking $k-m = j$, we get

$$\underline{n}_j C_n \subseteq C_{n-j} .$$

By considering the action of vector fields on linear functions, we find that

$$L_k = \sum_{m=1}^{r} \oplus\, C_{m-k} D_m$$

The decomposition $C_n = H_n \oplus C_{n+1}$ then gives (c), finishing the proof.

Associated to the filtered algebra $L = L_r \supset L_{r-1} \supset \cdots$ is the graded algebra

$$\mathrm{gr}(L) = \sum_{k=1}^{r} \oplus (L_k / L_{k-1}) .$$

The Lie algebra structure on $\mathrm{gr}(L)$ is defined as follows: If $T \in L_j$, $S \in L_k$, and $(T)_j$, $(S)_k$ denote their cosets mod L_{j-1} and L_{k-1} respectively, then

$$[(T)_j, (S)_k] = ([T,S])_{j+k} .$$

This is well-defined by virtue of the filtration condition, and obviously satisfies the Jacobi identity. From part (c) of the Proposition we have

Corollary $\quad \mathrm{gr}(L) \approx \underline{n}$.

This corollary can be viewed as the statement that near 0 , the infinite-dimensional Lie algebra L of all vector fields can be "approximated" by the finite-dimensional nilpotent Lie algebra $\underline{n}$. When the gradation on V is trivial ($r = 1$) , this approximation merely consists in replacing a vector field by its value at 0 . For $r > 1$, this approximation is more subtle, as it distinguishes among the various directions in V . We will return to this approximation question in chapter II, in connection with subalgebras of L .

Remark The gradation $\underline{n} = \underline{n}_1 \oplus \cdots \oplus \underline{n}_r$ can also be described using the dilations δ_t :

$$\underline{n}_k = \{T \in \underline{n} : T(f \circ \delta_t) = t^k (Tf) \circ \delta_t\} .$$

This follows immediately from the formula for $\underline{n}_k$. Thus the elements of $\underline{n}_k$ are "homogeneous of degree k" as vector fields.

1.4 Locally unipotent automorphisms Let Aut $(\mathcal{P})$ be the group of all automorphisms of the algebra $\mathcal{P}$. Set

$$N = \{\varphi \in \text{Aut } \mathcal{P} : (\varphi - I)\, \mathcal{P}_n \subseteq \mathcal{P}_{n-1}, \forall n\} .$$

It is clear that N is a subgroup of Aut $\mathcal{P}$. If $X \in \underline{n}$, $f \in \mathcal{P}$, define

$$e^X f = \Sigma \frac{1}{n!} X^n f .$$

This is a finite sum, by the local nilpotence of X.

Theorem. If $X \in \underline{n}$, then $e^X \in N$, and the map $X \longmapsto e^X$ is a bijection from $\underline{n}$ onto N.

Proof. Leibnitz' formula asserts that

$$X^n(fg) = \sum_k \binom{n}{k} X^k f\, X^{n-k} g .$$

Substituting this into the series definition of $e^X(fg)$ and rearranging, we conclude that e^X is a homomorphism. Since $e^X e^{-X} = I$, it follows that $e^X \in N$.

Conversely, given $\varphi \in N$, define a linear transformation X on $\mathcal{P}$ by

$$Xf = \sum_{n \geq 1} \frac{(-1)^{n+1}}{n} (\varphi - I)^n f .$$

Clearly $X \mathcal{P}_n \subseteq \mathcal{P}_{n-1}$, and by the formal power series identity $t = e^{\log t}$, we have $e^X = \varphi$. It only remains to prove that X is a derivation.

First we prove that $e^{tX} \in$ Aut $(\mathcal{P})$ for all $t \in \mathbb{R}$. Indeed, for any $n \in \mathbb{Z}$, $e^{nX} = \varphi^n$. Hence for any $f,g \in \mathcal{P}$, the function

$$t \longmapsto e^{tX}(fg) - (e^{tX} f)(e^{tX} g)$$

vanishes on $\mathbb{Z}$. But this is clearly a polynomial in t, so it must vanish identically.

Thus

$$X(fg) = \frac{d}{dt}\Big|_{t=o} e^{tX} (fg)$$

$$= \frac{d}{dt}\Big|_{t=o} (e^{tX}f)(e^{tX}g)$$

$$= (Xf)g + f\,(Xg) \,, \qquad \text{Q.E.D.}$$

The map $X \longmapsto e^X$ furnishes global "canonical coordinates of the first kind" for N. Using this map, we transfer the analytic manifold structure of the vector space $\underline{n}$ to the group N. If $X,Y \in \underline{n}$ and $f \in P$, then the map

$$X,Y \longmapsto e^X e^Y f$$

is obviously a polynomial mapping on $\underline{n} \times \underline{n}$. Since elements of N are determined by their restriction to $V^* \subseteq P_r$, it follows that group multiplication is a polynomial map when expressed in canonical coordinates. Indeed, as in the proof of the theorem, if $e^X e^Y = e^Z$, then

$$Z = \log (e^X e^Y) \in \underline{n} \,.$$

Hence if we write $Z = X * Y$, then

$$(*) \qquad X * Y = \sum_{n \geq 1} \frac{(-1)^{n+1}}{n} (e^X e^Y - I)^n \,.$$

To determine the expression for $X * Y$ as a vector field, we only need calculate $\{X * Y(\xi_i)\}$, using $(*)$. Since this series is locally finite on P, we find that $\{X * Y(\xi_i)\}$ are polynomial functions of $\{X(\xi_i) , Y(\xi_i)\}$. (In § 2 we shall obtain more explicit information about these functions using the Campbell-Hausdorff formula to rewrite $(*)$ in terms of Lie polynomials.)

1.5 <u>Transformation groups</u> The group N also has a dual presentation as a group of (non-linear) analytic transformations of the vector space V. Let $\{\xi_i\}$ be a basis for V^* with ξ_i of weight n_i

<u>Theorem</u> If $\varphi \in N$, then there is a transformation $T : V \to V$ of the form

$$(*) \qquad \xi_i(Tx) = \xi_i(x) + q_i(x) ,$$

with $q_i \in P_{n_i-1}$, such that $\varphi(f)(x) = f(Tx)$. Conversely, for any choice of $q_i \in P_{n_i-1}$, formula $(*)$ defines an analytic isomorphism of V, and there exists $\varphi \in N$ such that

$$\varphi(f) = f \circ T , \qquad \text{for all } f \in P .$$

<u>Proof</u> Since $\varphi(\xi_i) = \xi_i \mod P_{n_i-1}$, there exist q_i so that $\varphi(\xi_i) = \xi_i + q_i$. Define T by $(*)$. Then since P is generated by V^*, it is clear that $\varphi(f) = f \circ T$.

Conversely, given any $q_i \in P_{n_i-1}$, there exists a unique <u>homomorphism</u> $\varphi : P \to P$ such that $\varphi(\xi_i) = \xi_i + q_i$. One has $(\varphi - I) P_n \subseteq P_{n-1}$, so the same proof as in § 1.4 shows that $\varphi = e^X$ for some $X \in \underline{n}$. In particular, φ is invertible, with $\varphi^{-1} = e^{-X}$. Hence $\varphi \in N$. Clearly $\varphi(f) = f \circ T$, so that if $S : V \to V$ is the transformation corresponding to φ^{-1}, we have $S \circ T = T \circ S = I$, Q.E.D.

1.6 <u>Finite-dimensional representations</u> We have presented the group N as a subgroup of Aut (P) and as a subgroup of the group of analytic isomorphisms of the manifold V, two infinite-dimensional groups. We may also embed N as a subgroup of a linear group.

Let $\pi_n : N \to GL(P_n)$ be the finite-dimensional representation of N obtained by restriction to P_n. Since $\pi_n(\varphi) f = f \mod P_{k-1}$ for $f \in P_k$, the matrix for $\pi_n(\varphi)$, relative to the decomposition $P_n = H_0 \oplus H_1 \oplus \cdots \oplus H_n$, is

$$\pi_n(\varphi) \longleftrightarrow \begin{pmatrix} I_1 & * & \cdots & * \\ & I_2 & & \vdots \\ 0 & & \ddots & * \\ & & & I_n \end{pmatrix},$$

where I_k = identity transformation on H_k .

<u>Theorem</u> If $n \geq r$, the representation π_n is faithful.

<u>Proof</u> If $\varphi \in \mathrm{Ker}\,(\pi_n)$ and $n \geq r$, then $\pi_{|V^*} = I$. Since V^* generates P as an algebra, this implies that $\varphi = I$.

1.7 <u>Examples</u> 1) If $V = V_1$, then δ_t is scalar multiplication by t , H_n is the usual space of homogeneous polynomials of degree n , $\underline{n}$ = all constant-coefficient vector fields, and N = all translations by elements of V .

2) If $V = V_1 \oplus V_2$, then $H = V_1^*$ and $H_2 = (V_1^*)^2 + V_2^*$. If $\underline{n}_k$ is defined as in § 1.3 , then $\underline{n}_1 = \mathcal{D}_1 \oplus V_1^* \mathcal{D}_2$ and $\underline{n}_2 = \mathcal{D}_2$. Suppose $\dim V_1 = \dim V_2 = 1$, and (x,y) are coordinate functions. Then $\underline{n}_1$ is spanned by $X = \partial/\partial x$ and $Y = x\partial/\partial y$, and $\underline{n}_2$ is spanned by $\partial/\partial y = [X,Y]$. Thus $\underline{n}$ is the three-dimensional Heidenberg algebra in this case. As a transformation group on $\mathbb{R}^2$, N acts by

$$\begin{cases} x \to x + t_1 & , \quad t_i \in \mathbb{R} \\ y \to y + t_2\, x + t_3 \end{cases}$$

The space P_2 has basis 1 , x , y , x^2 , and the faithful representation π_2 maps N onto matrices

$$\begin{pmatrix} 1 & t_1 & t_3 & t_1^2 \\ 0 & 1 & t_2 & 2t_1 \\ 0 & 0 & 1 & 0 \\ 0 & 0 & 0 & 1 \end{pmatrix}, \qquad t_i \in \mathbb{R}$$

§ 2. Birkhoff Embedding Theorem

2.1 Filtrations on nilpotent Lie algebras Let $\underline{g}$ be a finite-dimensional Lie algebra over $\mathbb{R}$. A positive filtration F of $\underline{g}$ is a chain of subspaces

$$\underline{g}_1 \supseteq \underline{g}_2 \supseteq \underline{g}_3 \supseteq \cdots$$

such that

$$\begin{cases} \underline{g} = \underline{g}_1 , \quad \underline{g}_n = 0 \quad \text{for } n \text{ large} \\ [\underline{g}_j , \underline{g}_k] \subseteq \underline{g}_{j+k} . \end{cases}$$

Proposition $\underline{g}$ is nilpotent $\iff$ there exists a positive filtration of $\underline{g}$.

Proof Set $\underline{g}^1 = \underline{g}$, $\underline{g}^{n+1} = [\underline{g} , \underline{g}^n]$, so that $\{\underline{g}^n\}$ is the descending central series. By definition, $\underline{g}$ is nilpotent if an only if $\underline{g}^n = 0$ for n large. Thus we must show that $[\underline{g}^m , \underline{g}^n] \subseteq \underline{g}^{m+n}$. For $m = 1$ this is true, by definition, for all n. Assume that for some m it is true for all n. Then

$$[\underline{g}^{m+1} , g^n] = [[\underline{g} , \underline{g}^m] , \underline{g}^n]$$

by Jacobi: $\quad \subseteq [\underline{g} , [\underline{g}^m , \underline{g}^n]] + [\underline{g}^m , [\underline{g} , \underline{g}^n]]$

by induction: $\quad \subseteq [\underline{g} , \underline{g}^{m+n}] + [\underline{g}^m , \underline{g}^{n+1}]$

$$\subseteq \underline{g}^{m+n+1} .$$

Conversely, if $\{\underline{g}_n\}$ is any positive filtration, then $\underline{g}^1 = \underline{g}_1$, so $\underline{g}^2 = [\underline{g}_1 , \underline{g}_1] \subseteq \underline{g}_2$, and inductively, $\underline{g}^n \subseteq \underline{g}_n$. Hence $\underline{g}^n = 0$ for n large, and $\underline{g}$ is nilpotent, Q.E.D.

Fix a positive filtration $F = \{\underline{g}_n\}$ on $\underline{g}$. For each n, choose a linear subspace V_n so that $\underline{g}_n = V_n \oplus \underline{g}_{n+1}$. Then as a vector space,

$$(*) \qquad \underline{g} = V_1 \oplus \cdots \oplus V_r ,$$

where r is the length of the filtration. Define dilations δ_t, homogeneous norm $|x|$, and the spaces H_n, P_n of polynomials on $\underline{g}$ as in § 1, relative to this decomposition. Denote by $N(F)$ the nilpotent group constructed in § 1 . By the remark in § 1.1 , the spaces P_n , and hence the group $N(F)$, only depend on the filtration, and not on the choice of complementary subspaces V_n .

Definition The decomposition (*) is a gradation of $\underline{g}$ if $[V_j, V_k] \subseteq V_{j+k}$ (Set $V_n = 0$ for $n > r$). Equivalently, the dilations δ_t associated with the decomposition are automorphisms of the Lie algebra $\underline{g}$ in this case.

2.2 Algebraic comparison of additive and nilpotent group structures Given the nilpotent Lie algebra $\underline{g}$, we define a Lie group structure on $\underline{g}$ by the Campbell-Hausdorff formula. Recall that the formula asserts that, e.g. if one considers the algebra of formal power series in two non-commuting indeterminants X, Y, then

$$(\#) \qquad e^X e^Y = e^{F(X,Y)} ,$$

where

$$F(X,Y) = X + Y + \frac{1}{2}[X,Y] + \frac{1}{12}[X,[X,Y]] + \frac{1}{12}[Y,[Y,X]] + \cdots$$

is a (universal) formal Lie series in X,Y .

Write

$$F(X,Y) = X + Y + \tau(X,Y) .$$

By the identity (#) one sees that τ has no terms in X alone or in Y alone.

Now for $x,y \in \underline{g}$, define $\tau(x,y)$ by substitution in the formal Lie series $\tau(X,Y)$. Since $\underline{g}$ is nilpotent, this defines a polynomial map $\tau : \underline{g} \times \underline{g} \to \underline{g}$, with $\tau(0,y) = \tau(x,0) = 0$. Set

$$xy = x + y + \tau(x,y) .$$

By the formal identity (#) it follows that this composition gives a group structure to $\underline{g}$. The straight line $t \longmapsto tx$, $t \in \mathbb{R}$, is the one-parameter group generated by x . Denote by G the space $\underline{g}$ with this Lie group structure.

To study the algebraic structure of the map τ , we dualize to obtain a comultiplication on the algebra P of polynomial functions on $\underline{g}$. We make the canonical identification of $P \otimes P$ with the polynomial functions on $\underline{g} \oplus \underline{g}$. Then we can define an algebra homomorphism

$$\mu : P \to P \otimes P$$
$$\mu(f)(x,y) = f(xy) .$$

Similarly, using the additive structure of $\underline{g}$, we define

$$\Delta : P \to P \otimes P$$
$$\Delta(f)(x,y) = f(x+y) .$$

We want to compare the homomorphisms μ and Δ . (If $[\underline{g},\underline{g}] = 0$, then $\mu = \Delta$.)

The filtration $\{P_n\}$ of P induces canonically a filtration on $P \otimes P$, by

$$(P \otimes P)_n = \sum_{i+j \leq n} P_i \otimes P_j .$$

Write $Q_n = (P \otimes P)_n$, and let

$$Q_n^+ = \{h \in Q_n : h(x,0) = h(0,y) = 0\} .$$

Theorem If $f \in P_n$, then $\mu(f) = \Delta(f) \mod Q_n^+$.

Proof It is clear that $Q_m Q_n^+ \subseteq Q_{m+n}^+$. If $\xi \in \underline{g}^*$, then $\Delta(\xi) = \xi \otimes 1 + 1 \otimes \xi$. Since Δ is a homomorphism, this implies that $\Delta(P_n) \subseteq Q_n$. From these two properties we conclude that the subspace of P for which the theorem is true is a subalgebra. Thus we only need verify the theorem when f is a linear function.

Let $\xi \in \underline{g}^*$. Then $\mu(\xi) = \Delta(\xi) + \xi \circ \tau$. But τ is a Lie polynomial, without any linear terms.

Lemma If $c(x,y)$ is any Lie polynomial without linear terms, then there exist polynomials $q_k \in Q_k^+$ and elements $z_k \in \underline{g}_k$ such that

$$c(x,y) = \sum_{k\geq 2} q_k(x,y) z_k .$$

Proof of Lemma Write $x = \Sigma \xi_i(x)x_i$, $y = \Sigma \xi_i(y)x_i$, where $\{x_i\}$, $\{\xi_i\}$ are dual bases for $\underline{g}$, $\underline{g}^*$, with $x_i \in V_{n_i}$, $\xi_i \in V_{n_i}^*$.

Then

$$[x,y] = \Sigma \xi_i(x)\xi_j(y) z_{ij} ,$$

where $z_{ij} = [x_i,x_j] \in \underline{g}_{n_i+n_j}$ and $\xi_i \otimes \xi_j \in Q_{n_i+n_j}^+$. The Lemma follows by induction on the length of the commutators in c .

Applying the lemma to τ , we see that if $\xi \in V_n^*$, then $\xi \circ \tau \in Q_n^+$, since $\xi(\underline{g}_k) = 0$ when $k > n$. This proves the theorem.

Remarks 1. Let $\{\xi_i\}$ be a basis for $\underline{g}^*$ with $\xi_i \in V_{n_i}$. The statement of the theorem is equivalent to the following statement about the formula for xy in canonical coordinates:

Set $a_i = \xi_i(x)$, $b_i = \xi_i(y)$, $c_i = \xi_i(xy)$.

Then

$$c_i = a_i + b_i + q_i(a_1,\ldots,a_{i-1}, b_1,\ldots,b_{i-1}) ,$$

where $q_i \in Q_{n_i}^+$ (Since q_i contains no terms in $\{a_j\}$ alone or $\{b_j\}$ alone, it depends only on the a_j,b_j with $n_j < n_i$.)

From this formula it is clear, for example, that Euclidean measure on $\underline{g}$ is invariant under right and left translation by elements of G , since the Jacobian determinant is 1 .

2. Suppose $\underline{g} = V_1 \oplus \cdots \oplus V_r$ is a gradation of $\underline{g}$. Set $K_n = \Sigma H_i \otimes H_j$ $(i + j = n)$, so that K_n consists of the homogeneous polynomials of weight n on $\underline{g} \oplus \underline{g}$, relative to the dilations $(\delta_t x , \delta_t y)$. Let K_n^+ be the same sum,

but with $i \geq 1$, $j \geq 1$ and $i + j = n$. Then

$$\mu(f) = \Delta(f) \mod K_n^+ , \quad \forall f \in H_n .$$

Indeed, δ_t is an automorphism in this case, so $\tau(\delta_t x , \delta_t y) = \delta_t \tau(x,y)$. Thus this property holds for linear functions f , and hence for all polynomials, by the same argument as before, with Q_n , Q_n^+ replaced by K_n , K_n^+ .

2.3 Faithful unipotent representations Let $R, L : G \to \text{Aut}(P)$ be the right and left regular representations of G on P :

$$R(x) f(y) = f(yx) , L(x) f(y) = f(x^{-1}y) .$$

Both these representations are faithful, since the functions in P separate the points of G .

Theorem R and L map G into the group $N(F)$ associated with the filtration F of $\underline{g}$.

Proof Let $\rho_x : P \otimes P \to P$ be the homomorphism defined by "evaluation on the right at x":

$$\rho_x(F)(y) = F(y,x)$$

Then $R_x = \rho_x \circ \mu$. Also $x \to \rho_x \circ \Delta$ is simply the regular representation of the additive group of $\underline{g}$ on P , which clearly is in $N(F)$.

Now Q_n^+ is spanned by monomials $\xi^\alpha \otimes \xi^\beta$,with $w(\alpha) + w(\beta) \leq n$ and $w(\alpha) \geq 1$, $w(\beta) \geq 1$. In particular, $w(\alpha) \leq n-1$, so that $\rho_x (Q_n^+) \subseteq P_{n-1}$. Combining these facts, we see that when $f \in P_n$,

$$\begin{aligned} R_x f &= \rho_x \mu(f) \\ &= \rho_x \Delta(f) \mod \rho_x (Q_n^+) \\ &= f \mod P_{n-1} \end{aligned}$$

(we have used the theorem of § 2.2 to pass from μ to Δ) .

This proves that $R_x \in N(F)$. To pass to left translation, let $J \in \mathrm{Aut}\,(P)$ be defined by $Jf(x) = f(-x)$. Then $JP_n = P_n$ and $JR_xJ = L_x$ (Note that $-x$ is the inverse to x in G). Hence $L_x \in N(F)$, Q.E.D.

<u>Corollary 1</u> If dR is the differential of R:

$$dR(x)\ f(y) = \frac{d}{dt}\Big|_{t=0} f(y(tx)) \quad ,$$

then dR is a Lie algebra homomorphism from $\underline{g}$ to $\underline{n}$.

<u>Corollary 2</u> If $U_n(x) = R(x)|P_n$, then U_n is a finite-dimensional unipotent representation of G . If the filtration F is of length r , then U_n is faithful when $n \geq r$.

Both these corollaries follow immediately from the theorem and § 1.4 - 1.6 .

<u>Remark</u> Suppose $\underline{g} = V_1 \oplus \cdots \oplus V_r$ is a <u>gradation</u> of $\underline{g}$. Let $\underline{n} = \underline{n}_1 \oplus \cdots \oplus \underline{n}_r$ be the gradation of $\underline{n}$ defined in § 1.3 . Then

$$x \in V_k \Longrightarrow dR(x) \in \underline{n}_k \ .$$

This is easily proved directly from the fact that δ_t is an automorphism in this case, using the remark at the end of § 1.3 .

§ 3. <u>Comparison of group structures</u>

3.1 <u>Norm comparison of additive and nilpotent structures</u> In this section we want to convert the algebraic information in Theorem 2.2 into an estimate for the difference between the additive and nilpotent group structures on the nilpotent Lie algebra $\underline{g}$. Let F be a filtration on $\underline{g}$, and choose a homogeneous norm on $\underline{g}$ compatible with the decomposition (*) in § 2.1 .

<u>Theorem</u> There is a constant $C > 0$ such that

(†) $|xy-x-y| \leq C\{|x|^a|y|^{1-a} + |x|^a|y|^a + |x|^{1-a}|y|^a\}$, where $a = 1/r$, r = length of filtration Γ .

<u>Proof</u> Since all homogeneous norms are equivalent, we may assume that

$$|x| = \max |\xi_i(x)|^{1/n_i} ,$$

where $\{\xi_i\}$ is a basis for V^* with $\xi_i \in V^*_{n_i}$. Then

$$\xi_i(xy-x-y) = p_i(x,y) ,$$

where $p_i \in Q^+_{n_i}$, by Theorem 2.2 . But p_i is a sum of monomials $\xi^\alpha(x)\xi^\beta(y)$, with $w(\alpha) + w(\beta) \leq n_i$ and $w(\alpha) \geq 1$, $w(\beta) \geq 1$. Since $|\xi^\alpha(x)| \leq |x|^{w(\alpha)}$, we thus have

$$|p_i(x,y)| \leq C \max \{|x|^j|y|^k\} ,$$

where $1 \leq j, k$ and $j + k \leq n_i$. From this we obtain the estimate

$$|xy-x-y| \leq C \max \{|x|^{j/n}|y|^{k/n}\} ,$$

where the max is taken over all integers j,k,n with $j \geq 1$, $k \geq 1$, $2 \leq n \leq r$, and $j + k \leq n$.

To find the dominant term in this estimate, assume $|y| \geq |x| > 0$ and write

$$|x|^{j/n}|y|^{k/n} = \left(\frac{|x|}{|y|}\right)^{j/n} |y|^{(j+k)/n} .$$

We may assume $r \geq 2$, since otherwise $[\underline{g},\underline{g}] = 0$ and $xy = x+y$. Thus $a \leq 1/2$, and for j,k,n in the indicated range we have

$$j/n \geq a , \quad 2a \leq (j+k)/n \leq 1 .$$

Hence

$$|x|^{j/n}|y|^{k/n} \leq \left(\frac{|x|}{|y|}\right)^a \max \{|y|^{2a} , |y|\} .$$

Interchanging x and y, we get estimate $(\dagger)$

Corollary Suppose the filtration F comes from a gradation of $\underline{g}$, and $|x|$ is a homogeneous norm relative to the gradation. Then

$$(\dagger\dagger) \qquad |xy-x-y| \leq C\{|x|^a|y|^{1-a} + |x|^{1-a}|y|^a\},$$

where $a = 1/r$, r = length of F.

Proof of Corollary Let ξ_i, p_i be as in the proof above. Since δ_t is an automorphism of the Lie algebra in the graded case, we have

$$p_i(\delta_t x, \delta_t y) = t^{n_i} p_i(x,y)$$

Hence p_i is a sum of monomials $\xi^\alpha(x)\xi^\beta(y)$ with $w(\alpha) + w(\beta) = n_i$, $w(\alpha) \geq 1$, $w(\beta) \geq 1$. By the proof just given, this leads to the same estimates as before, but now with the constraint $j + k = n$. Hence the term $|x|^a|y|^a$ does not occur in the final estimate.

Remark The only difference between the filtered and the graded case in these estimates is the behaviour near $x = 0$, $y = 0$. As long as either $|x| \geq \varepsilon > 0$ or $|y| \geq \varepsilon > 0$, estimates $(\dagger)$ and $(\dagger\dagger)$ are equivalent.

3.2 Algebraic comparison of filtered and graded structures Let $F = \{\underline{g}_n\}$ be a decreasing filtration of the nilpotent algebra $\underline{g}$. We construct a graded nilpotent Lie algebra $gr(\underline{g})$ using F, as follows: Set

$$gr(\underline{g}) = \sum_{n\geq 1} \oplus (\underline{g}_n / \underline{g}_{n+1}),$$

and define

$$[X + \underline{g}_{m+1}, Y + \underline{g}_{n+1}] = [X,Y] + \underline{g}_{m+n+1},$$

when $X \in \underline{g}_m$, $Y \in \underline{g}_n$. The right-hand side of this formula only depends on the equivalence classes of X, Y mod $\underline{g}_{m+1}$, $\underline{g}_{n+1}$, respectively, by virtue of the filtration condition. Extending this bracket operation to a bilinear map of $gr(\underline{g})$ into $gr(\underline{g})$, we obtain a Lie algebra structure (skew-symmetry and the Jacobi identity follow immediately from the corresponding identities in $\underline{g}$).

In this section we want to make an algebraic comparison between the Lie algebras $\underline{g}$ and $gr(\underline{g})$. Pick a linear map $\alpha : \underline{g} \to gr\, \underline{g}$ such that for all n,

$$\alpha(X) = X + \underline{g}_{n+1} \quad , \quad \text{if } X \in \underline{g}_n \ .$$

Then α is a linear isomorphism, and we transfer the Lie multiplication from $\underline{g}$ to $gr(\underline{g})$ by defining

$$\mu(x,y) = \alpha\left(\left[\alpha^{-1}x, \alpha^{-1}y\right]\right) \ .$$

If we denote $\underline{g}_n / \underline{g}_{n+1} = V_n$, $gr(\underline{g}) = V$, then we see from the filtration property $\left[\underline{g}_n, \underline{g}_m\right] \subseteq \underline{g}_{m+n}$ that the bilinear map μ can be written as a finite sum of bilinear maps

$$\text{(1)} \qquad \mu = \mu_o + \mu_1 + \ldots + \mu_{r-1} \ ,$$

where

$$\mu_k : V_m \times V_n \to V_{m+n+k}$$

(r = length of the filtration). In particular, μ_o is the Lie algebra multiplication on V defined above. Each of the maps μ_k is skew symmetric.

Define dilations δ_t on V by

$$\delta_t\, x = t^n x, \quad x \in V_n \ .$$

If $b : V \times V \to V$ is a bilinear map, define

$$\delta_t^* b(x,y) = \delta_{1/t}\, b(\delta_t\, x, \delta_t\, y) \ .$$

The maps μ_k are thus homogeneous of degree k : $\delta_t^* \mu_k = t^k \mu_k$. Thus

$$\delta_t^* \mu = \mu_o + t\mu_1 + \ldots + t^{r-1} \mu_{r-1} \ .$$

In particular,

$$\lim_{t\to o} \delta_t^* \mu = \mu_o .$$

Note that for every $t \neq o$, $\delta_t^* \mu$ defines a Lie algebra multiplication on V, and the Lie algebra $(V, \delta_t^* \mu)$ is isomorphic to $\underline{g}$, via the linear map $\delta_t \circ \alpha$. Thus $gr(\underline{g})$ is in the <u>closure</u> of the isomorphism class of $\underline{g}$.

When is $gr(\underline{g})$ actually <u>isomorphic</u> to $\underline{g}$? This will occur exactly when we can choose the map α above so that $\mu_k = 0$ for $k \geq 1$. But the possible choices are of the form $\varphi\alpha$, where $\varphi : V \to V$ is linear and

$$\varphi\big|_{V_n} = \text{Identity} \pmod{\sum_{k>n} V_k} .$$

If we transfer the Lie multiplication from $\underline{g}$ to V using $\varphi\alpha$, we obtain a bilinear map ν on V such that

$$\varphi\, \mu(x,y) = \nu(\varphi x, \varphi y) .$$

As before, we decompose ν into its homogeneous parts:

$$\nu = \nu_o + \nu_1 + \ldots + \nu_{r-1} , \tag{2}$$

where $\delta_t^* \nu_k = t^k \nu_k$.

We have $\nu_o = \mu_o$, since this gives the Lie multiplication of $gr(\underline{g})$.

To compare (1) and (2), we note that φ can be written as

$$\varphi = I + \varphi_1 + \ldots + \varphi_{r-1} ,$$

where

$$\varphi_k : V_n \to V_{n+k} .$$

Hence equating terms of the same degree of homogeneity (relative to δ_t^*) gives the relations

$$\sum_{m+n=p} \varphi_m \mu_n(x,y) = \sum_{i+j+k=p} \nu_k(\varphi_i x, \varphi_j y) , \tag{3}$$

for $0 \leq p \leq r-1$ (φ_0 = Identity) . In particular, $\underline{g} \simeq gr(\underline{g})$ if and only if we can pick φ_k so that $\nu_k = 0$ for $k \geq 1$.

To express these equations in a more informative way, we introduce the coboundary operator associated with the Lie algebra $gr(\underline{g})$. Let $C^n(V,V)$ be the space of alternating, n-linear maps from V to V . Define

$$\delta : C^n(V,V) \to C^{n+1}(V,V)$$

by the formula

$$\delta f(x_1,\ldots,x_{n+1}) = \sum_{i<j} (-1)^{i+j} f(\mu_o(x_i,x_j), x_1,\ldots,\hat{x}_i,\ldots,\hat{x}_j,\ldots,x_{n+1})$$

$$- \sum (-1)^i A(x_i)f(x_1,\ldots,\hat{x}_i,\ldots,x_{n+1}) .$$

Here $\hat{x}_i$ means to omit x_i , and A is the adjoint representation of $gr(\underline{g})$, i.e. $A(x)y = \mu_o(x,y)$. The Jacobi identity for μ_o is equivalent to the condition $\delta^2 = 0$. For $n = 1$ the formula for δf becomes

$$\delta f(x_1,x_2) = \mu_o(f(x_1), x_2) + \mu_o(x_1,f(x_2))$$

$$- f(\mu_o(x_1,x_2)) .$$

Using this, we obtain from (3) the following criterion:

Proposition The Lie algebras $\underline{g}$ and gr $\underline{g}$ are isomorphic $\iff$ There exist linear maps φ_p on V , $\varphi_p : V_n \to V_{n+p}$, such that

$$\delta\varphi_p = \mu_p + F_p \quad , \quad 1 \leq p \leq r-1 \quad , \tag{4}$$

where $F_p \in C^2(V,V)$ is defined by

$$F_p(x,y) = \sum_{k=1}^{p-1} \varphi_k\, \mu_{p-k}(x,y) - \mu_o(\varphi_k x, \varphi_{p-k} y)$$

$(F_1 = 0)$.

Remarks 1. The set of equations (4) seems quite intractable. There is additional information available, however, which we have not used; namely, that μ satisfies the Jacobi identity. This can be expressed most neatly by introducing the interior product of $f, g \in C^2(V,V)$: $f \cdot g \in C^3(V,V)$ and is defined by

$$(f\cdot g)(x_1,x_2,x_3) = f(g(x_1,x_2),x_3) + f(g(x_2,x_3),x_1) + f(g(x_3,x_1),x_2) .$$

The Jacobi identity is then $\mu\cdot\mu = 0$. Using the decomposition (1) of μ this gives the equations

$$(5) \quad \begin{cases} \mu_o \cdot \mu_o = 0 \\ \mu_o \cdot \mu_1 + \mu_1 \cdot \mu_o = 0 \\ \mu_o \cdot \mu_p + \mu_p \cdot \mu_o = -\sum_{k=1}^{p-1} \mu_k \cdot \mu_{p-k} . \end{cases}$$

We have already noted the first of these equations. As for the others, we calculate that

$$\mu_o \cdot \mu_p + \mu_p \cdot \mu_o = -\delta\mu_p$$

Hence μ_1 is a 2-cocycle $(\delta\mu_1 = 0)$, and the μ_p for $p > 1$ satisfy

$$\delta\mu_p = \sum_{k=1}^{p-1} \mu_k \cdot \mu_{p-k} .$$

We conclude that necessary conditions for solving (4) recursively for φ_p , starting with $p = 1$, are that the cohomology classes of $\mu_p + F_p$ be zero.

2. The filtration F determines an intrinsic cohomology class $[F] \in H^2(\text{gr } \underline{g}, A)$ (A = adjoint representation of gr $\underline{g}$) . Indeed by (3) ,

$$\mu_1(x,y) - \nu_1(x,y) = \mu_o(\varphi_1 x,y) + \mu_o(x,\varphi_1 y) - \varphi_1\mu_o(x,y) = \delta\varphi_1(x,y) ,$$

so that μ_1 and ν_1 are representatives of the same cohomology class. By remark 1 , this class is the "first obstruction" to constructing an isomorphism between $\underline{g}$ and $gr(\underline{g})$.

Examples 1. If the filtration is of length ≤ 2 , then $gr(\underline{g})$ is isomorphic to $\underline{g}$. Indeed, in this case $[\underline{g}_2,\underline{g}] = 0$, so if V_1 any complement to $\underline{g}_2$ in $\underline{g}$, then $[V_1,V_1] \subseteq \underline{g}_2$, $[V_1,\underline{g}_2] = 0$.

2. Consider the family $\underline{g}\,(a,b,c,d)$ of seven-dimensional nilpotent Lie algebras with basis $x_1,\ldots,x_7$ and commutation relations

$$\begin{cases} [x_1,x_n] = x_{n+1} , \quad [x_1,x_7] = 0 \\ [x_2,x_3] = ax_5 + bx_6 + cx_7 \\ [x_2,x_4] = ax_6 + bx_7 \\ [x_2,x_5] = (a-d)x_7 \\ [x_3,x_4] = d\,x_7 , \end{cases}$$

with all other commutators obtained by skew-symmetry from this table (or equal zero if they do not appear in the table). A straightforward calculation shows that Ad x_i is a derivation of this algebra structure, $1 \leq i \leq 7$, and hence these equations do define a four-parameter family of Lie algebras.

The descending central series $\underline{g}^n$ is given by

$$\underline{g}^n = \text{span}\,\{x_{n+1},\ldots,x_7\} ,$$

when $2 \leq n \leq 6$. Thus $\underline{g}$ is six-step nilpotent. It is obvious from the form of the commutation relations that $gr\,\underline{g} \cong \underline{g}\,(0,0,0,0)$, relative to the filtration $\{\underline{g}^n\}$. But clearly $\underline{g}\,(0,0,0,0)$ is not isomorphic to $\underline{g}\,(a,b,c,d)$ when any of the parameters are non-zero . Hence the descending central series is graded only when $\underline{g}$ is the semi-direct product of (x_1) and a six-dimensional commutative Lie algebra W , with ad x_1 being the shift operator on W .

A more interesting filtration on $\underline{g}$ is obtained by setting

$$\underline{g}_n = \text{span}\,\{x_n,x_{n+1},\ldots,x_7\} , \quad 1 \leq n \leq 7$$

This filtration is of length seven, and it is obvious that relative to this filtration

$$gr(\underline{g}) \approx \underline{g}\ (a,0,0,d) \quad ,$$

with the parameters a and d arbitrary. Letting $V_n = \text{span}\ (x_n)$, and writing $[x,y] = \mu(x,y)$, we see that the decomposition (1) of μ into a sum of homogeneous terms is given by

$$\begin{cases} \mu_o(x_1,x_n) = x_{n+1} \quad , \quad \mu_o(x_1,x_7) = 0 \\ \mu_o(x_2,x_3) = ax_5 \\ \mu_o(x_2,x_4) = ax_6 \\ \mu_o(x_2,x_5) = (a-d)x_7 \\ \mu_o(x_3,x_4) = dx_7 \end{cases}$$

$$\mu_1(x_2,x_3) = bx_6 \quad , \quad \mu_1(x_2,x_4) = bx_7$$

$$\mu_2(x_2,x_3) = cx_7 \quad , \quad \mu_3 = \ldots = \mu_6 = 0 \quad .$$

(All entries not obtainable by skew-symmetry are zero).

Let us examine equation (4) of the proposition in this case. A map φ_p , homogeneous of degree p , is defined by

$$\varphi_p\ x_n = a_n\ x_{n+p} \quad ,$$

where $a_n = 0$ when $n > 7-p$. The coboundary of φ_p , relative to the Lie algebra structure μ_o , is thus the skew-symmetric mapping defined by

$$\delta\varphi_p(x_i,x_j) = a_i\mu_o(x_{i+p}\ , x_j) + a_j\mu_o(x_i,x_{j+p}) - \varphi_p(\mu_o(x_i,x_j))$$

When $p = 1$ we calculate that

$$\begin{cases} \delta\varphi_1(x_1,x_2) = (a_2-a_3)\ x_4 \\ \delta\varphi_1(x_1,x_3) = (aa_1+a_3-a_4)\ x_5 \\ \delta\varphi_1(x_1,x_4) = (aa_1+a_4-a_5)\ x_6 \\ \delta\varphi_1(x_1,x_5) = ((a-d)a_1+a_5-a_6)\ x_7 \\ \delta\varphi_1(x_2,x_3) = a(a_3-a_5)\ x_6 \\ \delta\varphi_1(x_2,x_4) = (da_2+(a-d)a_4-aa_6)\ x_7\ , \end{cases}$$

with all other entries not obtainable by skew symmetry equal zero. Hence the condition $\delta\varphi_1 = \mu_1$ is equivalent to the set of 6 linear equations

$$\begin{cases} a_2-a_3 = 0 \\ a_3-a_4 = -aa_1 \\ a_4-a_5 = -aa_1 \\ a_5-a_6 = (d-a)\ a_1 \\ a(a_3-a_5) = b \\ d(a_2-a_4) + a(a_4-a_6) = b \end{cases}$$

One calculates that this inhomogeneous system has rank 5 , and is consistent if and only if either $b = 0$ (i.e. $\mu_1 = 0$) , or else $a \neq 0$. In particular, if $a = 0$ and $b \neq 0$, then we cannot solve $\delta\varphi_1 = \mu_1$, and hence the filtration is not equivalent to a gradation on $\underline{g}(0,b,c,d)$, $b \neq 0$.

Assume $a \neq 0$. Then there is no first-order obstruction to defining an isomorphism between $\underline{g}(a,b,c,d)$ and $\underline{g}(a,0,0,d,)$. The higher-order obstructions require solving $\delta\varphi_p = \mu_p + F_p$, when $p = 2,3,\ldots,6$. By calculations similar to those just made, we find that for $p > 2$ these equations can always be solved, while for $p = 2$ they can always be solved if $d \neq 0$. If $d = 0$, then using the explicit form of equation (4) and the solution φ_1 , we find that the consistency condition is expressed by the equation $5b^2 = 4ac$. Hence we conclude that <u>if</u> $a \neq 0$ <u>and</u> $d \neq 0$, <u>then</u> $\underline{g}(a,b,c,d) \cong \underline{g}(a,0,0,d)$.

Furthermore, $\underline{g}(a,b,c,0) \cong \underline{g}(a,0,0,0)$ if and only if $5b^2 = 4ac$. Finally, the same analysis shows that $\underline{g}(0,0,c,0) \cong \underline{g}(0,0,0,0)$ if and only if $c = 0$. (This is also obvious by inspection of the multiplication table in this case).

3.3 Norm comparison of filtered and graded structures Continuing the notation of the previous section, let us turn now to the question of a metric comparison between the group structures defined by $\underline{g}$ and $gr(\underline{g})$. Fix a linear map $\alpha : \underline{g} \to gr(\underline{g})$ such that $gr(\alpha) = I$ as in § 3.2 , and identify $\underline{g}_n$ with the subspace $\alpha(\underline{g}_n)$. Thus we have two nilpotent Lie algebra structures on the vector space V , corresponding to the Lie brackets μ and μ_0 . We shall write xy and $x*y$, respectively, for the corresponding Campbell-Hausdorff group laws on V . Fix a dilation-homogeneous norm $|x|$ on V . Then the maps $x,y \longmapsto xy$ and $x,y \longmapsto x*y$, are "asymptotic at infinity" when measured by the homogeneous norm, in the following sense (Recall that $\underline{g} = gr\ \underline{g}$ if the filtration is of length ≤ 2) :

Theorem Assume the filtration F is of length $r \geq 3$, and set $a = 1/r$. Then there is a constant M so that

$$|xy-x*y| \leq M\,(|x|^{1-2a}|y|^{a} + |x|^{a}|y|^{a} + |x|^{a}|y|^{1-2a}) \quad .$$

In particular, $|xy-x*y| \leq M\,(|x| + |y|)^{1-a}$ if $|x| + |y| \geq 1$. Thus

$$\lim_{|x|+|y|\to\infty} \frac{|xy-x*y|}{|x|+|y|} = 0 \quad .$$

Proof Since the Campbell-Hausdorff multiplication is given by a universal formula, it will suffice to compare the result of evaluating a formal Lie polynomial at $x,y \in V$, using the two Lie algebra structures on V .

Pick a basis $\{x_i\}$ for V and dual basis $\{\xi_i\}$ for V^* , with $x_i \in V_{n_i}$ and $\xi_i \in V^*_{n_i}$. We can write, by equation (1) of § 3.2 ,

$$\mu = \mu_o + \beta \quad ,$$

where $$\beta(V_m, V_n) \subseteq \underline{g}_{m+n+1} \quad .$$

Hence for the formal Lie element $c(x,y) = [x,y]$ we have

$$(*) \qquad \mu(x,y) = \mu_o(x,y) + \Sigma\ \xi_i(x)\xi_j(y)\ z_{ij} \quad ,$$

where $z_{ij} = \beta(x_i, x_j) \in \underline{g}_{n_i+n_j+1}$. More generally, for any Lie polynomial c , let us write $c(x,y)$ for the result of substitution using the Lie bracket μ , and write $c^*(x,y)$ for the result of substitution using the Lie bracket μ_o . Then by equation $(*)$ and induction one sees that

$$c(x,y) = c^*(x,y) + \sum_{k\geq 2} q_k\ (x,y)\ z_k \quad ,$$

where $q_k \in Q_k^+$ and $z_k \in \underline{g}_{k+1}$ (notation of § 2.2). Thus it suffices to estimate the maps $x,y \longmapsto q_k(x,y)\ z_k$.

As in the proof of Theorem 3.1 , we have the estimate

$$|q_k(x,y)| \leq C \max\ \{|x|^i |y|^j\} \quad ,$$

where $1 \leq i,j$ and $i+j \leq k$. Since $z_k \in \underline{g}_{k+1}$, it follows that

$$|q_k(x,y)\ z_k| \leq C \max\ \{|x|^{i/n}\ |y|^{j/n}\} \quad ,$$

where the max is taken over $k+1 \leq n \leq r$, with i,j as before. To find the dominant term in this estimate, we assume $|y| \geq |x| > 0$, and write

$$|x|^{i/n}\ |y|^{j/n} = \left(\frac{|x|}{|y|}\right)^{i/n} |y|^{(i+j)/n} \quad .$$

In the indicated range we have $i/n \geq a$ and $2a \leq (i+j)/n \leq 1-a$. Hence

$$|q_k(x,y)\ z_k| \leq C \left(\frac{|x|}{|y|}\right)^a \left(|y|^{2a} + |y|^{1-a}\right) \quad .$$

Symmetrising this estimate with respect to x and y, we obtain the first estimate of the theorem.

To obtain the second estimate, we note that

$$(|x||y|)^a \leq (|x| + |y|)^{2a}$$

$$|x|^{1-2a} |y|^a \leq (|x| + |y|)^{1-a} .$$

Since $a \leq 1/3$, one has $2a \leq 1-a$, so that for $|x| + |y| \geq 1$, $(|x| + |y|)^{2a} \leq (|x| + |y|)^{1-a}$. This gives the second estimate and completes the proof of the theorem.

<u>Corollary</u> Let $\{\delta_t\}$ be the dilations associated with $gr(\underline{g})$. Then the group structure $x*y$ of $gr(\underline{g})$ is obtained from the group structure xy of $\underline{g}$ by

$$x*y = \lim_{t\to\infty} \delta_{1/t} ((\delta_t x)(\delta_t y)) .$$

<u>Proof</u> Set $x_t = \delta_t x$, $y_t = \delta_t y$. Since δ_t is an automorphism of $gr(\underline{g})$, we have $x*y = \delta_{1/t} (x_t * y_t)$. Hence by the theorem

$$\begin{aligned} |x*y - \delta_{1/t} (x_t y_t)| &= t^{-1} |x_t * y_t - x_t y_t| \\ &\leq M\, t^{-a} (|x|^{1-2a}|y|^a + |x|^a |y|^{1-2a}) \\ &\quad + M\, t^{2a-1} |x||y| . \end{aligned}$$

But $x*y = xy$ if $a \geq 1/2$, so we may assume $a \leq 1/3$. Then this estimate gives

$$|x*y - \delta_{1/t} (x_t y_t)| = O(t^{-a})$$

when $t \to \infty$, proving the corollary.

<u>Remarks</u> 1. This Corollary makes more precise the statement above that the group structures defined by $\underline{g}$ and $gr\ \underline{g}$ are "asymptotic at infinity". Namely,

to calculate $x*y$, we dilate x and y by δ_t , t very large, form the product $x_t y_t$, and then shrink back with $\delta_{1/t}$. This gives an approximation to $x*y$ with an error $0(t^{-a})$.

2. All the estimates of this section are monotone <u>increasing</u> functions of the length r of the filtration $(a = 1/r)$, as long as we stay away from $x = 0, y = 0$.

Comments and references for Chapter I

§ 1.1-1.2 The notion of "homogeneous norm" (=gauge) and dilations was systematically developed in Knapp-Stein [1], and then in Korányi-Vági [1]. We have followed Korányi-Vági in requiring that the norm be homogeneous of degree one relative to dilations. For many applications the smoothness and symmetry of the norm are irrelevant, so we have not included these conditions as part of the definition. (In many estimates, norms like $|x|_\infty$ are the most convenient to use.)

§ 1.3 The Lie algebra of vector fields with polynomial coefficients has been studied, e.g. in Auslander-Brezin-Sacksteder [1], Arnol'd [1], and Goodman [6]. (cf. Appendix). The use of the homogeneous norm to define the order of vanishing of functions at a point is taken from Folland-Stein [1] and Rothschild-Stein [1]. The Corollary appears to be new.

§ 2.1 One of the earliest uses of the filtration given by the descending central series of a nilpotent Lie algebra is in Birkhoff [1], where it is extended to a filtration of the universal enveloping algebra by ideals J_n of finite codimension. The process of using a filtration on $\underline{g}$ to define filtrations on algebraic objects functorially attached to $\underline{g}$ has been systematically developed by Vergne [1] and Rauch [1], [2]. In these notes we restrict our attention to <u>integral</u> filtrations, because these suffice for the study of iterated commutators of vector fields. For the study of arbitrary positive filtrations, cf. Goodman [6]. We also restrict our attention to <u>one</u>-parameter groups of diagonalizable automorphism $\{\delta_t\}$. For the case of an n-parameter commutative group of diagonalizable automorphisms of a nilpotent Lie algebra, cf. Favre [1].

§ 2.2 The theory of free Lie algebras and the Campbell-Hausdorff formula can be found, e.g. in Bourbaki [1], Hochschild [1], and Jacobson [1].

§ 2.3 The construction given here is the dual to the construction of Birkhoff [1]. The linear dual space to the enveloping algebra of $\underline{g}$ is canonically isomorphic to the formal power series functions on $\underline{g}$ (cf. Dixmier [23]), and the annihilator of the ideal J_{n+1} of Birkhoff is the space P_n of polynomials of degree $\leq n$. For the analogue of the Birkhoff construction in the case of a solvable Lie algebra, cf. Reed [1].

§ 3.1 The comparison theorem was proved in Goodman [6]. It implies, in particular, that $|xy| \leq a\,(|x| + |y| + b)$, for some positive constants a and b. Similar estimates were proved by Guivarc'h [1], where it is shown that there exists a norm of the type $|x|_\infty$ for which the constant $a = 1$. We do not know if there is a smooth norm which satisfies $|xy| \leq |x| + |y| + b$. See also Jenkins [1].

§ 3.2 The main source for the results of this section is the thesis of Vergne [1]. (cf. Rauch [1], [2] for further developments). For the general theory of deformations of algebraic structures, cf. Nijenhuis-Richardson [1]; for Lie algebra cohomology, cf. Jacobson [1]. An example of a nilpotent Lie algebra with every automorphism unipotent was given by Dyer [1]. This phenomenon was investigated further by Müller-Römer [1], who showed that the 7 dimensional algebra $\underline{g}\,(0,1,0,1)$ of Example 2 has this property, but that every nilpotent Lie algebra of dimension ≤ 6 admits dilating automorphisms. The fact that the cohomology class of μ_1 is almost always trivial in Example 2 contrasts with Dixmier's results [1]: For any nilpotent $\underline{g}$, one has $\dim H^n(\underline{g},\underline{g}) \geq 2$ for $1 \leq n \leq \dim \underline{g} - 1$. R.W. Johnson [1] has studied the problem of constructing a gradation for the descending central series.

§ 3.3 The comparison theorem between the graded and filtered group structures was proved in Goodman [6]. The Corollary is due to Auslander-Brezin-Sacksteder [1], in a slightly different formulation.

Chapter II

Nilpotent Lie algebras as tangent spaces

§ 1. Transitive Lie algebras of vector fields

1.1 Geometric background In this chapter M will be a real C^∞ manifold. We denote by $C^\infty_{\mathbb{R}}(M)$ the algebra of real-valued smooth functions on M, and by TM the real tangent bundle to M. We let $L(M)$ be the Lie algebra of all real C^∞ vector fields on M. If $x \in M$, TM_x will denote the tangent space at x.

In this chapter we want to construct a theory of manifolds "modeled on nilpotent Lie groups". The motivation for such a theory comes from several sources (see Chapter IV). The basic geometric idea, however, is very simple. We would like to carry out in each vector space TM_x the constructions of Chapter I, and do this in a way that varies smoothly with x. This would mean putting the structure of a graded nilpotent Lie algebra on each tangent space TM_x, together with a Lie algebra homomorphism from TM_x to $L(M)$. (Here we are thinking of the group manifold, on which each tangent space is isomorphic via translations to $\underline{g}$, and $\underline{g}$ acts as vector fields by the regular representation.)

Thus this construction would involve splitting each tangent space into a direct sum of subspaces, assigning a degree of homogeneity to each subspace and commutation relations among the subspaces, respecting these degrees. We would then like to use analysis on the nilpotent groups at each tangent space (convolution operators, etc.) to carry out local analysis on M. When the splitting of the tangent spaces is trivial, so that the nilpotent groups are just the additive groups of the vector spaces TM_x, this is the essence of the "parametrix method", which has been enormously developed and refined in recent years (the theory of pseudo-differential operators and their symbolic calculus).

Despite the fact that graded nilpotent Lie algebras exist in great profusion, the problem of constructing a "bundle of nilpotent structures" as posed above seems hopelessly overdetermined, at first examination. To recast the problem into more reasonable form, let us note the following:

1) We want to avoid topological obstructions, so we are only working locally on M.

2) For purposes of local analysis, it is enough to model M on a homogeneous space for a nilpotent group, since functions on a homogeneous space can always be lifted to the group (This avoids the obvious problem of finding a graded nilpotent Lie algebra of the same dimension as M)

3) Since we want a bundle structure, we should first choose a fiber, namely a fixed, graded nilpotent Lie algebra $\underline{g}$. The data should then be the algebra $\underline{g}$ together with a map $\lambda : \underline{g} \to L(M)$. Denote by G the corresponding nilpotent group.

4) It is unreasonable to require that λ be an exact Lie algebra homomorphism, since this is unstable under small perturbations (see the example to follow). Instead, we should only require that λ be a "partial homomorphism".

5) The condition (2) above, stated in terms of λ, is simply that at each point $x \in M$, $\lambda(\underline{g})_x$ should be the full tangent space at x.

Now that we have relaxed the requirements on the data, there exist many such pairs $(\underline{g},\lambda)$. In fact, given any finitely generated Lie subalgebra $\underline{h}$ of $L(M)$ which is transitive in the sense that $\underline{h}_x = TM_x$ for $x \in M$, we can construct $(\underline{g},\lambda)$ with the additional condition that a set of generators for $\underline{g}$ maps onto a set of generators for $\underline{h}$. (This follows from the existence of free nilpotent algebras).

Suppose now that we have a pair $(\underline{g},\lambda)$ satisfying 3), 4), 5) (the precise definition of "partial homomorphism" will be given in the next section). The next step, in viewing M as an "approximate" homogeneous space for G, is to lift functions from M to G. This is no problem, at least locally. Identifying G with $\underline{g}$ as manifolds as always, we can lift a function f on M to a function $\tilde{f}$ on a neighborhood Ω of 0 in $\underline{g}$ by setting

$$\tilde{f}(u) = f(e^{\lambda(u)}x)$$

Here $e^{\lambda(u)}$ is the local flow on M generated by the vector field $\lambda(u)$, and x is a fixed point of M. This defines a map $W : f \to \tilde{f}$, which is injective (the map $u \to e^{\lambda(u)}x$ is surjective, by (5)).

If λ were a full Lie algebra homomorphism, then W would intertwine λ and the regular representation of $\underline{g}$. Since we only assume that λ is a partial homomorphism, this is not exactly true. The <u>lifting problem</u>, whose solution forms the core of this chapter, is to find a partial homomorphism

$$\Lambda : \underline{g} \to L(\Omega)$$

satisfying the transitivity condition (5), such that

(i) $\Lambda(Y)W = W\lambda(Y)$

(ii) $\Lambda(Y)$ is "well-approximated" by the left-invariant vector field on G determined by Y.

Once the lifting problem is solved, we can use the intertwining map W to transfer problems concerning the action of the vector fields $\lambda(Y)$ to problems about the "almost invariant" vector fields $\Lambda(Y)$.

Before stating and proving the "lifting theorem", we must make precise the notion of "partial homomorphism" and the approximation involved in (ii) above.

1.2 <u>Partial homomorphisms</u> To motivate the definition, consider the following example:

Let $X = \partial/\partial x$, $Y = x\,\partial/\partial y$, acting on $\mathbb{R}^2$. Away from the line $x = 0$, this pair of vector fields spans the tangent space. On the line $x = 0$, the vector field Y vanishes. However, $[X,Y] = \partial/\partial y$, so the Lie algebra generated by X and Y acts transitively, in the infinitesimal sense. Furthermore, if we set $Z = [X,Y]$, then X,Y,Z exactly span a nilpotent Lie algebra. If $\underline{g}$ is the Heisenberg Lie algebra, with basis ξ,η,ζ satisfying $[\xi,\eta] = \zeta$, then the map $\lambda : \xi \to X$, $\eta \to Y$, $\zeta \to Z$ extends to an (exact) homomorphism from $\underline{g}$ to vector fields on $\mathbb{R}^2$, and λ is transitive.

Suppose we perturb this example by replacing $x\, \partial/\partial y$ by $Y' \equiv \varphi(x)\, \partial/\partial y$, where φ is a smooth function with $\varphi(o) = 0$, $\varphi'(o) = 1$. Then the vector fields X, Y' and $Z' \equiv [X, Y']$ still span the tangent space at each point. However, the Lie algebra generated by X and Y' is generically neither nilpotent nor finite-dimensional (if $\varphi^{(n)}(o) \neq 0$ for infinitely many n) . On the other hand, it is only the subspace spanned by X, Y' and Z' which is geometrically important. If we let $\lambda' : \xi \to X$, $\eta \to Y'$, $\zeta \to Z'$, then λ' extends to a linear map from $\underline{g}$ into $L(\mathbb{R}^2)$ such that $\lambda'([\xi,\eta]) = [\lambda'(\xi), \lambda'(\eta)]$ (but in general, $\lambda'([\xi,\zeta]) \neq$ $\neq [\lambda'(\xi), \lambda'(\zeta)]$) .

Let us turn to the general situation. Suppose

$$\underline{g} = \sum_{k=1}^{r} \oplus V_k \quad , \quad [V_j, V_k] \subseteq V_{j+k}$$

is a graded nilpotent Lie algebra, and $\underline{h}$ is any Lie algebra.

<u>Definition</u> A linear map $\lambda : \underline{g} \to \underline{h}$ is a <u>partial homomorphism</u> (relative to the graded structure of $\underline{g}$) , if

$$\lambda([x,y]) = [\lambda(x) , \lambda(y)]$$

whenever $x \in V_j$, $y \in V_k$, and $j + k \leq r$.

When $\underline{h} = L(M)$ is the Lie algebra of vector fields on M , we will say that the linear map λ is <u>surjective at</u> $x \in M$ if $\lambda(\underline{g})_x = TM_x$.

<u>Example</u> Suppose $X_1, \dots, X_n \in L(M)$ are given, and let $\underline{h}$ be the Lie subalgebra of $L(M)$ generated by the $\{X_k\}$. Then for every integer $r \geq 1$, there exists a canonical graded Lie algebra $\underline{g}$ of length r and a partial homomorphism $\lambda : \underline{g} \to \underline{h}$, as follows:

Let F be the free Lie algebra on n generators $y_1, \dots, y_n$, and let

$\underline{g} = F/F^{r+1}$, with $Y_k = y_k + F^{r+1}$. Then by definition, $\underline{g}$ is the free r-step nilpotent Lie algebra on n generators $Y_1,\ldots,Y_n$.

Proposition $\underline{g} = V_1 \oplus\cdots\oplus V_r$ is graded, with $V_1 = \text{span}\,\{Y_1,\ldots,Y_n\}$ and V_k = span of k-fold commutators of $\{Y_i\}$. There exists a unique partial homomorphism

$$\mu : \underline{g} \to F$$

such that
$$\mu(Y_i) = y_i \ .$$

(Thus we can split the exact sequence $0 \to F^{r+1} \to F \to \underline{g} \to 0$ by a partial homomorphism).

Proof Let $\{H_\alpha\}$ be a Hall basis for the algebra F . Each H_α is a formal Lie monomial whose length we denote by $|\alpha|$. Let $F_n = \text{span}\,\{H_\alpha : |\alpha| = n\}$, so that F_n is the subspace of n-fold commutators of the generators $\{y_i\}$. Then $F = \Sigma \oplus F_n$, and $F^{r+1} = \Sigma\{F_n : n \geq r+1\}$.

Set $Y_\alpha = H_\alpha(Y_1,\ldots,Y_n)$. Then $\{Y_\alpha : |\alpha| \leq r\}$ is a basis for $\underline{g}$, and

$$V_k = \text{span}\,\{Y_\alpha : |\alpha| = k\} \ .$$

Define

$$\mu(Y_\alpha) = H_\alpha \ .$$

Then $\mu(V_k) = F_k$ for $1 \leq k \leq r$, and it follows that μ is a partial homomorphism. Since $\{Y_i\}$ generate $\underline{g}$, μ is clearly unique. Q.E.D.

Let us return now to the vector fields $X_1,\ldots,X_n \in L(M)$. By the universal property of the free Lie algebra F , there exists a unique Lie algebra homomorphism

$$\varphi : F \to \underline{h} \ ,$$

such that
$$\varphi(y_i) = X_i \ .$$

Setting $\lambda = \varphi \circ \mu$, we thus obtain a partial homomorphism

$$\lambda : \underline{g} \to \underline{h} ,$$

such that
$$\lambda(Y_i) = X_i .$$

In terms of the Hall basis, λ is simply the result of "substituting X_i for Y_i" in H_α :

$$\lambda(Y_\alpha) = H_\alpha(X_1,\ldots,X_n) .$$

Hence λ has the additional property that

$$\lambda(V_k) = \text{span of k-fold commutators of } \{X_i\} .$$

In particular, <u>suppose that the</u> X_i <u>together with commutators of the</u> X_i <u>of length</u> $\leq r$ <u>suffice to span the tangent space at</u> $x \in M$. <u>Then</u> λ <u>is surjective at</u> x .

1.3 <u>Lifting Theorem</u> Fix a partial homomorphism λ from a graded nilpotent Lie algebra $\underline{g}$ into $L(M)$, and <u>assume that</u> λ <u>is surjective at some point</u> $x \in M$.

<u>Lemma</u> There exists an open set Ω around 0 in $\underline{g}$, and an open set M' around x in M , such that the map

$$u \to e^{\lambda(u)}x$$

is a submersion from Ω onto M' .

<u>Notation</u> If X is a vector field, then the local flow generated by X will be denoted by $t \to e^{tX}$. By definition, the induced action on functions is given by

$$(*) \qquad \frac{d}{dt} f(e^{tX}y) = (Xf)(e^{tX}y) .$$

<u>Proof of Lemma</u> The flow $u \to e^{\lambda(u)}x$ exists for u near 0 in $\underline{g}$ by the fundamental existence theorem for ordinary differential equations. The differential

of this mapping at $u = 0$ is the map $u \to \lambda(u)_x$. The implicit function theorem thus gives the existence of such an oper set Ω .

Now fix the point x , and replace M by M' (i.e. drop the prime). We define the <u>intertwining operator</u>

$$W : C^\infty(M) \to C^\infty(\Omega)$$

(**) $$Wf(u) = f(e^{\lambda(u)}x) \ .$$

Let $dR : \underline{g} \to L(\Omega)$ be the right regular representation (we give $\underline{g}$ the group structure determined by the Campbell-Hausdorff formula):

$$dR(y)\ \varphi(u) = \frac{d}{dt}\Big|_{t=0} \varphi(u \cdot ty)$$

The lifting theorem asserts that W "approximately" intertwines $\lambda(y)$ and $dR(y)$. (The choice of right instead of left actions is dictated by definition (*)). To make this precise, let

$$\underline{g} = V_1 \oplus \cdots \oplus V_r$$

be the decomposition of $\underline{g}$ as a graded Lie algebra. Viewing this as a decomposition of $\underline{g}$ as a vector space, we apply the constructions of Chapter I, § 1.3: Let

$$C = C^\infty(\Omega)$$

$$C_m = \{f \in C : f(u) = O(|u|^m) \text{ near } 0\} ,$$

where $|u|$ is a homogeneous norm on $\underline{g}$. Let $L = L(\Omega)$, and let L_k be the vector fields of order $\leq k$ at 0 .

With these preliminary notions established, we can now state the central result of this chapter:

<u>Lifting Theorem</u> If λ is a partial homomorphism from the graded Lie algebra $\underline{g}$ into the Lie algebra of vector fields on a manifold, which is surjective at some point x , then there exists a neighborhood Ω of 0 in $\underline{g}$ and a

linear map

$$\Lambda : \underline{g} \to L(\Omega) \quad ,$$

such that

(i) $W\lambda(Y) = \Lambda(Y)W \qquad (Y \in \underline{g})$

(ii) $\Lambda(Y) = dR(Y) \mod L_{k-1} \qquad (Y \in V_k)$

(iii) Λ is surjective at 0 .

Furthermore, if $\underline{g}$ is the free r-step nilpotent Lie algebra, then Λ can be chosen to be a partial homomorphism.

Remark We shall call Λ a lifting of λ . Note that condition (iii) implies that the map $Y \to \Lambda(Y)_0$ is a linear isomorphism from $\underline{g}$ to the tangent space at 0.

§ 2. Proof of the Lifting Theorem

2.1 Basic Lie formulae Let X and Y be elements of an associative algebra. Let $D = \operatorname{ad} X$ be the derivation $D(T) = XT-TX$. Then one has the following identities of formal series:

(I) $$\frac{d}{dt}\Big|_{t=0} e^{X+tY} = e^X E(X)Y$$

(II) $$e^X Y = \frac{d}{dt}\Big|_{t=0} e^{X+tB(X)Y}$$

Here

$$E(X) = \frac{1-e^{-D}}{D} = \sum_{k\geq 0} \frac{(-1)^k}{(k+1)!} D^k$$

and

$$B(X) = \frac{D}{1-e^{-D}} = \sum_{k\geq 0} \frac{1}{k!} b_k D^k$$

(b_k = Bernoulli numbers).

The meaning of (I) and (II) is that the corresponding graded identities, obtained by equating terms of the same degree in X, are valid for all n:

(I)' $$\frac{1}{n!}\sigma(X^nY) = \sum_{k+m=n} \frac{(-1)^k}{(k+1)!m!} X^m D^k(Y)$$

(II)' $$\frac{1}{n!} X^nY = \sum_{k+m=n} \frac{1}{k!m!} b_k \; \sigma(X^m D^k(Y))$$

Here σ is the symmetrisation operator:

$$\sigma(X^nY) = \frac{1}{n+1}(X^nY + X^{n-1}YX + \ldots + YX^n)$$

Since $B(X) = E(X)^{-1}$, it is clear that (I) is equivalent to (II). To prove (I), note that $R = L-D$, where R is right multiplication by X and L is left multiplication by X. Also $LD = DL$. Hence

$$\sigma(X^nY) = \frac{1}{n+1} \sum_{i+j=n} X^i(L-D)^j(Y)$$

$$= \frac{1}{n+1} \sum_{j\leq n} \sum_{k\leq j} \binom{j}{k} (-1)^k X^{n-k} D^k(Y)$$

$$= n! \sum_{k+m=n} \frac{(-1)^k}{(k+1)!m!} X^m D^k(Y) \quad ,$$

where we have used the basic binomial identity

$$\sum_{j=k}^{n} \binom{j}{k} = \binom{n+1}{k+1} \quad .$$

We shall also need the identity

(III) $$\frac{d}{dt}\bigg|_{t=0} e^{X+tY+tZ} = \frac{d}{dt}\bigg|_{t=0} e^{X+tY} + \frac{d}{dt}\bigg|_{t=0} e^{X+tZ} \quad .$$

In graded terms this is simply the identity

(III)' $$\sigma(X^n(Y+Z)) = \sigma(X^nY) + \sigma(X^nZ) \quad .$$

2.2 Left-invariant vector fields Our first application of the formulae just proved will be to calculate an explicit formula for $dR(Y)$, $Y \in \underline{g}$.

Let $\underline{n}$ be the Lie algebra of § I.1.3 associated with the graded vector space $\underline{g}$.

Let $u \in \underline{g}$. Then the "Bernoulli operator"

$$B(u) = \frac{adu}{1-e^{-adu}}$$

defines a linear transformation on $\underline{g}$ (which is a polynomial function of u , by the nilpotence of adu) .

Lemma If $v \in \underline{g}$, $f \in C^{\infty}(\Omega)$, $u \in \Omega$, then

$$(\#) \qquad dR(v)f(u) = \frac{d}{dt}\Big|_{t=0} f(u+tB(u)v) .$$

Remark Formula (#) asserts that at the point u , the vector field $dR(v)$ is the directional derivative in the direction $B(u)v$.

Proof of Lemma It suffices to prove the Lemma when f is a polynomial function. Set $X = dR(u)$, $Y = dR(v)$. Then $X,Y \in \underline{n}$, by Theorem I.2.3 . Since dR is a Lie algebra homomorphism, we have

$$dR(B(u)v) = B(X)Y .$$

Now view G as embedded in $N = \exp \underline{n}$, via the right regular representation. Then

$$f(uv) = (e^{Y}f)(u) = e^{X}e^{Y}f(0) .$$

Hence

$$dR(v)f(u) = \frac{d}{dt}\Big|_{t=0} e^{X}e^{tY}f(0)$$

$$= e^{X}Yf(0)$$

$$= \frac{d}{dt}\Big|_{t=0} e^{X+tB(X)Y} f(0)$$

$$= \frac{d}{dt}\Big|_{t=0} f(u+tB(u)v) \quad .$$

(Note that by the local nilpotence of $\underline{n}$, all series here are finite when f is a polynomial.)

Remark The formula (#) is valid in any Lie group, when u is near 0 at least, and gives the explicit formula for left-invariant vector fields in exponential coordinates. The proof is the same as above, but now working in the pseudogroup of germs of analytic transformations, instead of the group N.

2.3 Formal solution Let $Y \in \underline{g}$, and set $X = \lambda(Y)$. Given $f \in C^\infty(M)$, the first step will be to calculate the asymptotic expansion of the function $WXf(u)$ around $u = 0$. For this purpose, it suffices to carry out all calculations in the algebra of formal series $\Sigma\varphi_n$, where $\varphi_n \in C_n$ vanishes to order $\geq n$ at 0 (as measured by the homogeneous norm). An equality

$$\varphi = \Sigma\, \varphi_n \qquad (\varphi_n \in C_n)$$

for $\varphi \in C^\infty(\Omega)$ will simply mean that for every m,

$$\varphi - \sum_{k\leq m} \varphi_k \in C_{m+1} \quad .$$

By definition of the flow generated by a vector field, we have

$$\frac{d}{dt}\Big|_{t=0} Wf(tu) = (\lambda(u)f)(x) \quad .$$

Iterating this, we obtain the formal expansion

$$Wf(u) = \sum_{n\geq 0} \frac{1}{n!}\, \lambda(u)^n f(x) \quad .$$

Thus we can write $W = e^{\lambda(u)}$ in the sense of asympotic expansions. Hence by the basic Lie formula (II) ,

$$\text{(a)} \qquad WX = \frac{d}{dt}\Big|_{t=0} \exp\left[\lambda(u) + tB(\lambda(u))X\right] \ .$$

(This formula means that to calculate the Taylor series of WXf we can use the identities (II)' to express $\lambda(u)^n X$ in terms of $\sigma(\lambda(u)^m D_u^k(X))$, where $D_u = \mathrm{ad}\ \lambda(u)$. Similar "finite" interpretations can be given to the formulas which follow.)

Since λ is only a partial homomorphism , $B(\lambda(u))X \neq \lambda(B(u)Y)$. However, if $Y \in V_k$, then we find that

$$\text{(b)} \qquad B(\lambda(u))X = \lambda(B(u)Y) + \sum_{w(\alpha)+k>r} \xi^\alpha(u) Z_\alpha$$

where each Z_α is a vector field on M corresponding to a commutator of weight $>r$ of the $\lambda(Y_i)$, and $\{\xi_i\}$ is a graded basis for $\underline{g}^*$. Notice that this expansion converges in the asymptotic sense, since $B(u)Y$ is a polynomial function of u . Let us write this as

$$B(\lambda(u))X = \lambda(B(u)Y) + T_Y(u)$$

Substituting this in (a) and using the Lie formula (III) , we get

$$WX = \frac{d}{dt}\Big|_{t=0} \exp \lambda(u+tB(u)Y) + \frac{d}{dt}\Big|_{t=0} \exp\left[\lambda(u) + tT_Y(u)\right]$$

By Lemma II.2.2 and the formula $W = e^{\lambda(u)}$, the first term is $dR(Y)W$. Using this and the Lie formula (I) , we thus have

$$\text{(c)} \qquad WX = dR(Y)W + WE(\lambda(u))\ T_Y(u)$$

Suppose $Y \in V_k$, and consider the "error term" $E(\lambda(u))\ T_Y(u)$ in (c) . This is a formal sum of terms $\xi^\alpha(u)\ T_\alpha$, with $w(\alpha) > r-k$ and T_α a vector field on M . From the surjectivity hypothesis, we conclude that there exists a neighborhood of x in M on which every vector field can be expressed as a linear combination, with coefficients in $C^\infty(M)$, of vector fields $\{\lambda(Z) : Z \in \underline{g}\}$. Hence by shrinking M, we can write $E(\lambda(u))\, T_Y(u)$ as a series of terms $\xi^\alpha(u)\, \varphi\, \lambda(Z)$, with $w(\alpha) > r-k$, $\varphi \in C^\infty(M)$, and $Z \in \underline{g}$. Applying the operator W to such a term, we obtain a term

$$\psi\, W\, \lambda\,(Z)\ ,$$

where $\psi(u) = \xi^\alpha(u)\, W\varphi(u) \in C_{w(\alpha)}$.

Let $\{Y_i : 1 \le i \le d\}$ be a basis for $\underline{g}$, with $Y_i \in V_{n_i}$. By formula (c) and the foregoing analysis, we can find functions $\psi_{ij}^{(n)} \in C_n$ such that

$$\text{(d)} \qquad W\,\lambda(Y_i) = dR(Y_i)W + \sum_{n\ge 1} \sum_{j=1}^{d} \psi_{ij}^{(n)}\, W\,\lambda(Y_j)\ .$$

This series converges in the asymptotic sense. Furthermore, we know that

$$n \le r-n_i \Rightarrow \psi_{ij}^{(n)} = 0\ .$$

To write this formula in more compact form, introduce the column vectors

$$\vec{X} = (\lambda(Y_i))\ , \quad \vec{Y} = (dR(Y_i))$$

and the matrices

$$\Psi_n = (\psi_{ij}^{(n)})\ .$$

Define $W\vec{X} = (W\,\lambda(Y_i))$. Then (d) becomes

$$\text{(d)'} \qquad W\vec{X} = \vec{Y}\,W + \sum_{n\ge 1} \Psi_n\, W\vec{X}$$

To complete the formal solution, we introduce the matrix $S = \Sigma\, \Psi_n$. This series converges in the asymptotic sense, and S vanishes to order ≥ 1 at $u=0$. Hence the geometric series

$$T = \sum_{n\ge 1} S^n$$

converges in the asymptotic sense. Since

(d)' can be written as $(I-S)\, W \vec{X} = \vec{Y} W$, it follows that

(e) $$W \vec{X} = (I+T)\, \vec{Y} W .$$

2.4 C^∞ solution To pass from the formal asymptotic expansion (e) to a solution in terms of C^∞ vector fields, we invoke a classical theorem of E. Borel: Given a formal power series, there exists a C^∞ function having the given series as its Taylor series at 0. Hence there exists a matrix $T^\sim$, whose entries are C^∞ functions on Ω, such that $T^\sim = T$ in the sense of formal Taylor series at $u = 0$. By the definition of asymptotic equality, we can conclude from equation (e) that

(†) $$W \vec{X} = (I + \tilde{T})\, \vec{Y} W \mod C_\infty ,$$

where $C_\infty \equiv \bigcap_n C_n$ is the space of functions vanishing to infite order at $u = 0$.

The space C_∞ is invariant under arbitrary C^∞ changes of coordinates on Ω. (This is not true of the spaces C_n, even with respect to linear transformations.) By the implicit function theorem we can describe the range of the intertwining operator W as follows: There are coordinates $(t_1,\ldots,t_d)$ for Ω such that $W\, C^\infty(M)$ consists of the functions depending on the first m coordinates ($m = \dim M$).

Set $\vec{Z}_0 = (I+\tilde{T})\, \vec{Y}$. Then equation (†) together with the above description of the range of W implies that there exist column vectors $\vec{F}_k$ of functions in C_∞ such that

$$W \vec{X} = \vec{Z}_0 W + \sum_{k=1}^{m} \vec{F}_k \frac{\partial}{\partial t_k} W$$

Replace $\vec{Z}_o$ by $\vec{Z} = \vec{Z}_o + \Sigma F_k(\partial/\partial t_k)$. Then

$$(\#) \qquad W\vec{X} = \vec{Z}W .$$

If $\{Z_i\}$ are the components of $\vec{Z}$, then the assumption that λ is a partial homomorphism together with this intertwining property implies that

$$(\#\#) \qquad [Z_i, Z_j]\, W = W\, \lambda([Y_i, Y_j])$$

if $n_i + n_j \leq r$.

We shall construct the desired lifting Λ by using certain of the Z_i and certain commutators of the Z_i . Let us first observe that

$$(\#\#\#) \qquad Z_i = dR(Y_i) \quad \mathrm{mod} \quad L_{n_i-1}$$

Indeed, we only need to verify this for the formal solution $(I+T)\vec{Y}$, since this property only depends on the Taylor expansions at $u = 0$ of the coefficients of the Z_i . But T is a series of terms $\Psi_{m_1} \cdots \Psi_{m_k}$, $k \geq 1$, and one can write

$$(\Psi_{m_1} \cdots \Psi_{m_k} \vec{Y})_i = \Sigma\, \varphi_{ij}\, dR(Y_j) ,$$

with $\varphi_{ij} \in C_{k(r-n_i+1)}$, as is easily verified by induction (using the vanishing property of the $\Psi_{ij}^{(n)}$) . Since the vector fields $dR(Y_j)$ are all of order $\leq r$ at 0 , this shows that $(T\vec{Y})_i$ is of order $\leq n_i - 1$ at 0 .

If $\underline{g}$ is the free, r-step nilpotent Lie algebra with generators $Y_1,\ldots,Y_n$, then we define Λ as follows: Let F be the free Lie algebra on n generators $y_1,\ldots,y_n$, as in § 1.2 . By the universal property of F , there exists a unique Lie algebra homomorphism

$$\Gamma : F \to L(\Omega)$$

such that $\Gamma(y_i) = Z_i$, $1 \leq i \leq n$. Let $\mu : \underline{g} \to F$ be the partial homomorphism of Proposition 1.2, and define $\Lambda = \Gamma \circ \mu$. Then Λ is obviously a partial homomorphism. Note that in terms of a Hall basis, one has

$$\Lambda(H_\alpha(Y_1,\dots,Y_n)) = H_\alpha(Z_1,\dots,Z_n) , \quad |\alpha| \leq r .$$

It follows from (##) and (###) that Λ satisfies properties (i) and (ii) of the Theorem.

When $\underline{g}$ is not the free, r-step nilpotent Lie algebra, then we cannot define Λ merely by specifying the vector fields corresponding to a set of generators for $\underline{g}$. In this case, we set $\Lambda(Y_i) = Z_i$, $1 \leq i \leq d$, and extend Λ by linearity to $\underline{g}$. It then follows from (#) and (###) that Λ satisfies properties (i) and (ii) of the Theorem.

It only remains to verify that Λ is surjective at 0. But this is an immediate consequence of property (ii) and the structure of L_k given by Proposition I.1.3. We know that at 0, $dR(Y)_0 = D_Y$, by Lemma II.2.2. All the vector fields in L_0 vanish at 0. Hence if $\underline{n} = \underline{n}_1 \oplus \cdots \oplus \underline{n}_r$ is the Lie algebra of § I.1.3 associated with the direct sum decomposition $\underline{g} = V_1 \oplus \cdots \oplus V_r$, then for $Y \in V_n$

$$\text{(T)} \qquad \Lambda(Y)_0 = D_Y\big|_0 \mod \sum_{j<n} \underline{n}_j\big|_0$$

But by the structure of $\underline{n}_j$, we have $\underline{n}_j\big|_0 = \mathcal{D}_{V_j}$. Hence the system of equations (T) is in triangular form, which implies that

$$\dim \Lambda(\underline{g})_0 = \dim \underline{g} .$$

This completes the proof of the lifting theorem.

§ 3. Group germs generated by partial isomorphisms

3.1 Exponential coordinates. Let $\underline{g}$ be a graded nilpotent Lie algebra of length r, and let

$$\lambda : \underline{g} \to L(M)$$

be a partial homomorphism. We shall assume that for $x \in M$, the map

$$u \longmapsto \lambda(u)_x$$

is a linear isomorphism from $\underline{g}$ onto TM_x.

We define the exponential map at each point $x \in M$ by

$$\exp_x(u) = e^{\lambda(u)}x .$$

The map $\exp_x$ is a diffeomorphism from some open neighborhood Ω of zero in $\underline{g}$ onto an open neighborhood of x. The size of this neighborhood depends on Lipschitz constants for the coefficients of the vector fields $\lambda(u)$. Thus if we fix $x_0 \in M$, we can find a neighborhood M_0 of x_0 and a fixed $\Omega \subset \underline{g}$ such that

1) if $x \in M_0$, then $\exp_x$ is a diffeomorphism on Ω, and $\exp_x(\Omega) \supseteq M_0$.

2) the map $x,u \mapsto \exp_x(u)$ is C^∞ from $M_0 \times \Omega$ to M.

We now consider the "group germ" consisting of the transformations $\{\varphi_u : u \in \Omega\}$, where

$$\varphi_u(x) = \exp_x(u) .$$

Pick an open subset Ω_0 compactly contained in Ω, such that

$$u,v \in \Omega_0,\ x \in M_0 \Rightarrow e^{\lambda(u)}\, e^{\lambda(v)}\, x \in M_0 .$$

Given $x \in M_0$, it then follows by conditions 1) and 2) above that there exists a C^∞ map

$$F_x : \Omega_0 \times \Omega_0 \to \Omega$$

such that

$$e^{\lambda(v)} e^{\lambda(u)} x = \exp_x (F_x(u,v))$$

Furthermore, the map $x,u,v \to F_x(u,v)$ is smooth.

In the special case $M = G$, x = identity, and λ = regular representation of $\underline{g}$ as vector fields on G, then F_x is the nilpotent group multiplication on $\underline{g}$ defined by the Campell-Hausdorff formula.

3.2 Comparison of group germs. The comparison between the additive and nilpotent group structures on $\underline{g}$ in § I.3.1 extends to the present situation, as follows:

Theorem Let $a = 1/r$, and let $|u|$ be a homogeneous norm on $\underline{g}$. There exists a constant C such that

$$|F_x(u,v)-u-v| \leq C(|u|^a|v|^{1-a} + |v|^a|u|^{1-a})$$

for all $u,v \in \Omega_0$ and $x \in M_0$.

Proof Fix $x \in M_0$, and write $F = F_x$, $\exp = \exp_x$. Since $F(0,0) = 0$, the series expansion

$$f(\exp F(u,v)) = \Sigma \frac{1}{n!} \lambda(F(u,v))^n f(x)$$

is valid in the asymptotic sense, for any $f \in C^\infty(M)$. But by definition of $F(u,v)$, we also have an asympotic expansion

$$f(\exp F(u,v)) = \Sigma \frac{1}{m!n!} \lambda(u)^m \lambda(v)^n f(x) .$$

Hence these two asymptotic expansions must agree. We conclude that the vector field $\lambda(F(u,v))$ has an asymptotic expansion given by the power series for

$$\log (e^{\lambda(u)} e^{\lambda(v)}) .$$

(where $\log T = \sum_{n\geq o} \frac{(-1)^n}{n+1} (T-1)^n$ is the formal series inverse to the exponential series) .

By Dynkin's explication of the Campbell-Hausdorff formula, we know that the formal series

$$\log (e^X e^Y) = X + Y + \cdots$$

in the non-commuting indeterminants X,Y can be rearranged to be expressible entirely in terms of X,Y and iterated commutators of X and Y . Applying this rearrangement in the case $X = \lambda(u)$, $Y = \lambda(v)$ and using the fact that λ is a partial homomorphism, we conclude that the asymptotic expansion of $\lambda(F(u,v))$ has the form

$$\lambda(F(u,v)) = \lambda(uv) + R(u,v) .$$

Here uv is the nilpotent group product on $\underline{g}$, and the remainder R is the sum of terms

$$p(u,v) \varphi X_\alpha ,$$

where p is homogeneous of total degree $n > r$ in (u,v) , $\varphi \in C^\infty(M)$, and $\{X_\alpha\}$ is the image under λ of a graded basis for $\underline{g}$. (Because of the surjectivity of λ , all the commutators of $\lambda(u)$, $\lambda(v)$ which occur in R are in the $C^\infty(M)$ - module spanned by the $\{X_\alpha\}$.) Since λ is an injective linear map, this implies that the asymptotic expansion of $F(u,v)$ is of the form

$$(*) \qquad F(u,v) = uv + \sum_{n>r} F_n(u,v) ,$$

where F_n is a homogeneous polynomial of total degree n in (u,v) (relative to the graded structure on $\underline{g}$).

To obtain the estimate of the theorem, we recall from Corollary I.3.1 that this estimate is satisfied by the difference $uv - u - v$. Hence we only need to estimate $F(u,v) - uv$, as u and v range over the bounded set Ω_o.

We note that $F_n(o,v) = F_n(u,o) = 0$. Hence

$$\|F_n(u,v)\| \leq C \sum_{k=1}^{n-1} |u|^k |v|^{n-k}$$

Also, on compact subsets of $\underline{g}$ we have

$$|w| \leq C \|w\|^a ,$$

where $a = 1/r$ and $\|\cdot\|$ denotes a Euclidean norm on $\underline{g}$. Applying these two estimates to (✱), we conclude that for $u,v \in \Omega_o$,

$$|F(u,v) - uv| \leq C \max_{\substack{k+j=r+1 \\ 1 \leq k \leq r}} (|u|^{ak} |v|^{aj})$$

To majorize the right side, assume that $|u| \leq |v|$. Then

$$\begin{aligned} |u|^{ak} |v|^{aj} &= \left(\frac{|u|}{|v|}\right)^{ak} |v|^{1+a} \\ &\leq \left(\frac{|u|}{|v|}\right)^{a} |v|^{1+a} . \end{aligned}$$

Hence we get the estimate, for any $u,v \in \Omega_o$,

$$|F(u,v) - uv| \leq C(|u|^a |v| + |u| |v|^a)$$

Since $|u| \leq C|u|^{1-a}$ on Ω_o, this completes the proof of the theorem.

In the course of proving the theorem, we also established the following comparsion between the group germ F and the nilpotent group structure on $\underline{g}$:

<u>Corollary</u> Let uv be the nilpotent group multiplication on $\underline{g}$, and let $a = 1/r$ (r = length of $\underline{g}$). There exists a constant C such that

$$|F_X(u,v) - uv| \leq C(|u|^a |v| + |u| |v|^a)$$

for all $u,v \in \Omega_0$ and $x \in M_0$. In particular, $|F_x(u,v) - uv| \leq C(|u| + |v|)^{1+a}$, so that

$$\lim_{|u|+|v|\to o} \frac{|F_x(u,v)-uv|}{|u|+|v|} = 0 \quad .$$

Remark The corollary shows that the map $u,v \to F_x(u,v)$ is tangent to the map $u,v \to uv$ to order greater then one at $(0,0)$. In this situation we have perturbed the graded nilpotent structure by adding "higher order" terms. Compare this result with Theorem I.3.3, which treated the perturbation by "lower order" terms.

§ 4. Examples from complex analysis

4.1 Real hypersurfaces in $\mathbb{C}^{n+1}$ Some interesting examples of transitive Lie algebras of vector fields arise in connection with "analysis on the boundary" of domains in $\mathbb{C}^{n+1}$ $(n \geq 1)$. Here one finds a simple geometric relation between the "holomorphic flatness" of the boundary and the length of the nilpotent Lie algebras which occur.

Let $M \subset \mathbb{C}^{n+1}$ be a real, C^∞ submanifold of real dimension $2n+1$. Thus M is locally definable as the set $\{f = o\}$, where f is a real-valued C^∞ function with $df \neq 0$. All our considerations will be local, so we may assume, e.g. that $M = \{f = o\} \cap U$, where U is open in $\mathbb{C}^{n+1}$. Denote by $C^\infty(M,\mathbb{C})$ the algebra of complex-valued C^∞ functions on M, and let

$$L_{\mathbb{C}}(M) = \mathrm{Der}(C^\infty(M,\mathbb{C}))$$

denote the Lie algebra of vector fields on M with coefficients in $C^\infty(M,\mathbb{C})$. For any $T \in L_{\mathbb{C}}(M)$, $\overline{T}$ will denote the conjugate vector field

$$\overline{T}(f) = \overline{T(\overline{f})} \ ,$$

and we define $\mathrm{Re}\, T = (1/2)(T+\overline{T})$, $\mathrm{Im}\, T = (1/2i)(T-\overline{T})$.

Let A be the algebra of holomorphic functions on the ambient space $\mathbb{C}^{n+1}$. The inclusion map $M \subset \mathbb{C}^{n+1}$ induces the restriction map

$$A \to C^{\infty}(M,\mathbb{C}) \ .$$

We define the holomorphic vector fields on M to be those vector fields which annihilate the anti-holomorphic functions:

$$L^{1,0}(M) = \{X \in L_{\mathbb{C}}(M) \ : \ X\,\overline{f} = 0 \text{ for all } f \in A\} \ .$$

Similarly, the anti-holomorphic vector fields annihilate the holomorphic functions:

$$L^{0,1}(M) = \{X \in L_{\mathbb{C}}(M) \ : \ X\, f = 0 \text{ for all } f \in A\} \ .$$

Clearly

$$L^{1,0} = \overline{L^{0,1}} \ .$$

The following properties also hold:

Proposition $L^{1,0}$ and $L^{0,1}$ are Lie subalgebras of $L_{\mathbb{C}}(M)$, with

$$L^{1,0} \cap L^{0,1} = 0 \ .$$

At each point $p \in M$, $\dim_{\mathbb{C}} (L^{1,0})_p = n$, so that

$$(L^{1,0} + L^{0,1})_p$$

is a 2n-dimensional complex subspace of the complexified tangent space $T_{\mathbb{C}}M_p$.

Proof It is clear from the definition that $L^{1,0}$ and $L^{0,1}$ are Lie subalgebras of $L_{\mathbb{C}}(M)$. If $X \in L^{1,0} \cap L^{0,1}$, then X annihilates the real and imaginary parts of the complex coordinate functions, and hence $X = 0$. It remains to calculate $\dim_{\mathbb{C}} (L^{1,0})_p$.

If the equation $f = 0$ defines M near p, with $df \neq 0$, then after performing an affine transformation on $\mathbb{C}^{n+1}$, we can assume $p = o$ and

$$df_p = dz_{n+1} + d\bar{z}_{n+1} ,$$

where $\{z_k\}$ are the complex coordinates on $\mathbb{C}^{n+1}$. Let us call $z_{n+1} = w$. Then the vector fields

$$L_k = f_w \frac{\partial}{\partial z_k} - f_{z_k} \frac{\partial}{\partial w} , \quad 1 \leq k \leq n ,$$

are tangent to M, since $L_k \bar{f} = 0$.

Here

$$\frac{\partial}{\partial z_k} = \frac{1}{2} \left(\frac{\partial}{\partial x_k} - i \frac{\partial}{\partial y_k}\right) ,$$

where $z_k = x_k + iy_k$. Evidently $L_k \in L^{1,0}$, and at p,

$$(L_k)_p = \left(\frac{\partial}{\partial z_k}\right)_p ,$$

by the above normalization of f. This shows that

$$\dim_{\mathbb{C}}(L^{1,0})_p \geq n .$$

We must have equality, however, because

$$2 \dim_{\mathbb{C}}(L^{1,0})_p = \dim_{\mathbb{C}}(L^{1,0} + L^{0,1})_p$$

$$\leq \dim T_{\mathbb{C}}M_p = 2n+1 .$$

This completes the proof of the proposition.

4.2 Points of type m By Proposition 4.1, we see that the real and imaginary parts of the holomorphic vector fields on M span a $2n$ dimensional subspace of the $2n+1$ dimensional tangent space to M at each point. The question of interest now is to determine whether the missing direction can be obtained by taking iterated commutators of these vector fields. It is technically simpler and more natural to work in the complexified tangent space to M. Then it becomes a question of examining iterated commutators of holomorphic and anti-holomorphic vector fields on M.

Definition For each integer $m \geq 1$, let L_m be the module (over $C^\infty(M,\mathbb{C})$) of vector fields generated by Lie brackets of length $\leq m$ of the holomorphic and anti-holomorphic vector fields on M. A point $p \in M$ is of type m if

$$(L_{m+1})_p = T_{\mathbb{C}}M_p$$

and $(L_m)_p \neq T_{\mathbb{C}}M_p$. (If no such m exists, p is of infinite type.)

We shall say that $p \in M$ is of type $\leq m$ if it is of type s for some $s \leq m$. Then if p is of type m, it has a neighborhood consisting of points of type $\leq m$. Note that

$$L_1 = L^{1,0} + L^{0,1} ,$$

so $(L_1)_p$ is never the entire tangent space. Hence p is of type ≥ 1, necessarily. The notion of type is obviously invariant under holomorphic changes of local coordinates in the ambient space $\mathbb{C}^{n+1}$.

Proposition Suppose $p \in M$ is of type $m < \infty$. Let $Z_1,\ldots,Z_n$ be holomorphic vector fields on M which span $L^{1,0}$ at p, and let $\underline{g}$ be the free $m+1$ - step real nilpotent Lie algebra on $2n$ generators $X_1,\ldots X_n$, $Y_1,\ldots,Y_n$. Then there exists a partial homomorphism

$$\lambda : \underline{g} \to L(M)$$

such that $\lambda(X_j) = \operatorname{Re} Z_j$, $\lambda(Y_j) = \operatorname{Im} Z_j$, and

$$\lambda(\underline{g})_p = T\, M_p .$$

Proof. All that needs to be verified is the surjectivity condition. By assumption, $\{\mathrm{Re}\, Z_j\}$ and $\{\mathrm{Im}\, Z_j\}$ span L_1 (as a $C^\infty(M,\mathbb{C})$ module) near p. By induction we see that their commutators of length $\leq r$ span L_r around p. Since p is of type m, the span over $\mathbb{R}$ of commutators of length $\leq m+1$ is thus the full tangent space.

4.3 Geometric characterisation. The type of a point $p \in M$ is determined by the degree of "holomorphic flatness" of M at p, in the following way:

Theorem A point $p \in M$ is of type $m < \infty$ if and only if there is a complex submanifold of codimension one tangent to M at p to order m, but no codimension one complex submanifold tangent to a higher order at p.

As a preliminary to proving the theorem, we make an affine change of coordinates so that $p = o$ and the function f defining the boundary is given by

$$f = w + \overline{w} + \varphi \quad ,$$

where $(z_1,\ldots,z_n,w)$ are local holomorphic coordinates and $d\varphi_o = 0$. We will write $z = (z_1,\ldots,z_n)$ and

$$f_j = \partial f/\partial z_j \; , \quad f_{\overline{j}} = \partial f/\partial \overline{z}_j$$

$$f_w = \partial f/\partial w \; , \quad f_{\overline{w}} = \partial f/\partial \overline{w} \; .$$

The vector fields

$$L_k = f_w \frac{\partial}{\partial z_k} - f_k \frac{\partial}{\partial w} \; , \quad 1 \leq k \leq n \; ,$$

and their conjugates thus span L_1 around p. The vector field

$$T = f_{\overline{w}} \frac{\partial}{\partial w} - f_w \frac{\partial}{\partial \overline{w}}$$

together with L_1 spans the entire complexified tangent space to M around p. The type of p is the smallest integer m such that

$$T_p \in (L_{m+1})_p \quad .$$

With the local coordinates around p chosen as above, we can write any vector field X around p as

$$(*) \qquad X = c\frac{\partial}{\partial w} + d\frac{\partial}{\partial \bar{w}} + \sum_{i=1}^{n} a^i \frac{\partial}{\partial z_i} + b^i \frac{\partial}{\partial \bar{z}_i} \quad .$$

In particular, for the vector field $X = [L_j, \bar{L}_k]$, we calculate that

$$\begin{cases} a^i = \delta_{ij}\,(f_{\bar{k}}\, f_{w\bar{w}} - f_{\bar{w}}\, f_{\bar{k}w}) \\ b^i = -\delta_{ik}\,(f_j\, f_{w\bar{w}} - f_w\, f_{j\bar{w}}) \\ c = (f_{\bar{w}}\, f_{j\bar{k}} - f_{\bar{k}}\, f_{j\bar{w}}) \\ d = -(f_w\, f_{j\bar{k}} - f_j\, f_{\bar{k}w}) \quad . \end{cases}$$

Furthermore, under the action of ad L_j, the coefficients of X transform as follows:

$$\begin{cases} a^i \to \dfrac{\partial(a^i,f)}{\partial(z_j,w)} - \delta_{ij}\,\{cf_{ww} + df_{w\bar{w}} + \sum_{k=1}^{n} a^k f_{wk} + b^k f_{w\bar{k}}\} \\ b^i \to \dfrac{\partial(b^i,f)}{\partial(z_j,w)} \\ c \to \dfrac{\partial(c,f)}{\partial(z_j,w)} + cf_{jw} + df_{j\bar{w}} + \sum_{k=1}^{n} a^k f_{jk} + b^k f_{j\bar{k}} \\ d \to \dfrac{\partial(d,f)}{\partial(z_j,w)} \end{cases}$$

Using these formulae, it is a straightforward induction to establish the form of iterated commutators of the vector fields $L_k, \bar{L}_k$:

<u>Lemma 1</u> Suppose X is a sum of Lie monomials of length $m \geq 2$ in the

vector fields $\{L_k, \overline{L}_k \; ; \; 1 \leq k \leq n\}$. Then

1) The coefficients a^i, b^i, c, d in the expression (*) are sums of terms of the form

$$\pm D^1(f)\, D^2(f) \cdots D^m(f) \; ,$$

where each D^i is a differentiation to order d_i $(1 \leq d_i \leq m)$, and $d_1 + d_2 + \cdots + d_m = 2m-1$.

2) Each such term in a^i or b^i involves differentiation a total of m-1 times relative to $z, \overline{z}$ and a total of m times relative to $w, \overline{w}$.

3) Each such term in c or d involves differentiation a total of m times relative to $z, \overline{z}$, m-1 times relative to $w, \overline{w}$, and contains a factor of the form $D(f)$, where D is a differentiation in $z, \overline{z}$ only of order $\leq m$.

From Lemma 1, it is clear that if there exists a complex, codimension one submanifold tangent to M at p to order m , then

$$(L_m)_p \neq T_{\mathbb{C}} M_p \quad .$$

Indeed, in this case we can make a preliminary change of coordinates so that the hyperplane $\{w = o\}$ is tangent to M to order m at $p = o$. This simply means that $D(f) = o$ at o when D is any differentiation in $z, \overline{z}$ only of order $\leq m$. By part 3) of lemma 1 this implies that the $\partial/\partial w$ and $\partial/\partial \overline{w}$ coefficients of any $X \in L_m$ vanish at p . This proves that the type of p is at least as big as the maximal order of tangency of complex codimension one submanifolds.

To estimate the type of p from above, we first use the fomulas above and an induction argument to prove the following:

<u>Lemma 2</u>. Let $J, K \in \mathbb{N}^n$ be given, with $|J| \geq 1$, $|K| \geq 1$, and $|J| + |K| = m$. Then there exists a vector field X as in Lemma 1, such that

$$c = (f_w)^{|J|-1}(f_{\overline{w}})^{|K|}\left(\frac{\partial}{\partial z}\right)^J\left(\frac{\partial}{\partial \overline{z}}\right)^K f + R \quad ,$$

where R is a sum of terms of the form $D^1(f) \cdots D^m(f)$, and in each such term some D^i is a differentiation in $z,\overline{z}$ only of order $\leq m-1$.

With this lemma we can finish the proof of the theorem. Suppose p is of type $m < \infty$. By a holomorphic change of coordinates we can arrange that $\{w=o\}$ is tangent to M at p , and

$$\left(\frac{\partial}{\partial z}\right)^J f(p) = o \quad , \qquad |J| \leq m$$

$(z = (z_1,\ldots,z_n)$ as above) . Indeed, it suffices to replace w by $w' = w + p_m(z)$, where p_m is the holomorphic part of the Taylor polynomial for $f(z,o)$ of degree m . (Since $p_m(o) = o$, the functions $z_1,\ldots,z_n$, w' still form local holomorphic coordinates around p .) We are assuming that if $k \leq m$, then

$$(L_k)_p \neq T_{\mathbb{C}}M_p \quad .$$

By Lemma 2, this implies inductively that

$$\left(\frac{\partial}{\partial z}\right)^J \left(\frac{\partial}{\partial \overline{z}}\right)^K f(p) = o$$

for all J,K with $|J| + |K| \leq m$. Together with the above normalization of f , this shows that the complex hyperplane $\{w=o\}$ is tangent to M to order $\geq m$. This completes the proof of the theorem.

Remarks 1. Suppose the point $p \in M$ is of type $m < \infty$. The proof above shows that there are holomorphic coordinates w and $z = (z_1,\ldots,z_n)$ around p so that the Taylor series for f is of the form

$$(\dagger) \qquad f(z,w) = w + \overline{w} + \sum_{\substack{|\alpha|+|\beta|=r \\ \alpha\neq o,\beta\neq o}} C_{\alpha\beta}\, z^\alpha\, \overline{z}^\beta + \cdots \, .$$

Here $r = m + 1$, and the neglected terms are of the order $\|w\|^2 + \|w\| \|z\| + \|z\|^{2r+1}$. Furthermore, the constants $C_{\alpha\beta}$ are not all zero. This approximation for f fits naturally into the general framework of homogeneous norms. Namely, we define

$$|(z,w)| = (\|z\|^{2r} + \|w\|^2)^{1/2r} ,$$

where $\|z\|^2 = \Sigma \bar{z}_i z_i$, $\|w\|^2 = \bar{w}w$, and we define a filtration $\{C_k\}$ on the germs of C^∞ functions at p as in § 1.3, relative to this homogeneous norm. Then the error term in (†) is $O(|(z,w)|^{r+1})$, i.e. equation (†) is an equality modulo C_{r+1}.

2. By the remarks just made, we see that at a point of type m, the "first approximation" to the manifold M is the real-algebraic hypersurface

$$w + \bar{w} + \sum_{\substack{|\alpha|+|\beta|=r \\ \alpha\neq 0, \beta\neq 0}} C_{\alpha\beta}\, z^\alpha \bar{z}^\beta = 0$$

$(r = m + 1)$, where $\{C_{\alpha\beta}\} \neq 0$, and $C_{\alpha\beta} = \bar{C}_{\beta\alpha}$. In particular, when $m = 1$, $\{C_{ij}\}$ is a non-zero Hermitian matrix. In this case we can make a linear change of coordinates so that

$$f(z,w) = w + \bar{w} + \Sigma\, C_j \|z_j\|^2 + \cdots$$

where $C_j = 0$ or ± 1.

4.4 Siegel Domains and the Heisenberg group

Let us consider in more detail the example of the hypersurface

$$M = \{(z,w) : \mathrm{Im}(w) = \|z\|^2 , \quad z \in \mathbb{C}^n , w \in \mathbb{C}\} .$$

M is the boundary of the domain

$$D = \{(z,w) : \mathrm{Im}(w) > \|z\|^2\} ,$$

which is a Siegel domain of type II . By the foregoing analysis M is the simplest example of a "complex-convex" manifold of real codimension one. (Observe that the complex hyperplane $\{w = o\}$ is tangent to M to order one at 0 , and D lies strictly on one side of this hyperplane. As a consequence, the function $1/w$ is holomorphic in D and blows up exactly at the boundary point 0 .)

Since M is the zero set of the function $w - \overline{w} - 2i \,\|z\|^2$, the vector fields

$$L_j = \frac{\partial}{\partial z_j} + 2i\,\overline{z}_j \frac{\partial}{\partial w} \;, \quad 1 \le j \le n \;,$$

span the holomorphic tangent space everywhere on M , and mutually commute. The nontrivial commutation relations are

$$[L_j \,, \overline{L}_k] = \delta_{jk}\, N \;,$$

where $N = -2i\,(\frac{\partial}{\partial w} + \frac{\partial}{\partial \overline{w}})$. This shows that the complex Lie algebra generated by $\{L_j \,, \overline{L}_j : 1 \le j \le n\}$ is isomorphic to the complex $(2n+1)$ - dimensional Heisenberg algebra. As a basis for the real form of this algebra, we take

$$\begin{cases} X_j = \mathrm{Re}(L_j) = \frac{1}{2}\frac{\partial}{\partial x_j} + x_j \frac{\partial}{\partial t} + y_j \frac{\partial}{\partial s} \\ Y_j = \mathrm{Im}(L_j) = -\frac{1}{2}\frac{\partial}{\partial y_j} - y_j \frac{\partial}{\partial t} + x_j \frac{\partial}{\partial s} \\ Z = [X_j, Y_j] = \frac{\partial}{\partial s} \,, \end{cases}$$

where $z_j = x_j + iy_j$, $w = s + it$ in terms of real coordinates. Thus $\{X_j, \overline{Y}_j, Z \;;\; 1 \le j \le n\}$ span the real tangent space to M at each point, and satisfy the Heisenberg commutation relations

$$[X_j, Y_k] = \delta_{jk}\, Z$$

(all other commutators being zero).

Let $\underline{g}$ be the real Heisenberg algebra of dimension $2n+1$, and G the nilpotent Lie group pbtained from $\underline{g}$ as usual by the Campbell-Hausdorff formula.

Since $\underline{g}$ is two-step nilpotent, the group structure is

$$uv = u + v + \frac{1}{2}[u,v] , \qquad u,v \in \underline{g} .$$

Pick coordinates $\xi = (\xi_1,\ldots,\xi_n)$, $\eta = (\eta_1,\ldots,\eta_n)$ and ζ for $\underline{g}$ so that the map

$$\lambda(\xi,\eta,\zeta) = \sum_{j=1}^{n} (\xi_j X_j + \eta_j Y_j) + \zeta Z$$

is a Lie algebra isomorphism from $\underline{g}$ onto the Lie algebra spanned by $\{X_j, Y_j, Z\}$. For $u \in \underline{g}$, we shall write $u = (\xi,\eta,\zeta)$. Thus in these coordinates the commutation relations on $\underline{g}$ are

$$[(\xi,\eta,\zeta) , (\xi',\eta',\zeta')] = (0,0 , \xi\cdot\eta' - \xi'\cdot\eta) ,$$

where $\xi\cdot\eta = \Sigma\, \xi_j\eta_j$. Hence the group structure of G is

$$(\xi,\eta,\zeta)(\xi',\eta',\zeta') = (\xi+\xi',\eta+\eta', \zeta+\zeta' + \frac{1}{2}(\xi\cdot\eta' - \xi'\cdot\eta)).$$

Let us calculate the flow generated by the vector field $\lambda(\xi,\eta,\zeta)$, starting from the point 0 in $\mathbb{C}^{n+1}$. The integral curre

$$\tau \mapsto e^{\tau\lambda(\xi,\eta,\zeta)} \cdot 0$$

satisfies the system of ordinary differential equations

$$\begin{cases} \dfrac{dx_j}{d\tau} = \dfrac{1}{2}\xi_j , & \dfrac{dy_j}{d\tau} = -\dfrac{1}{2}\eta_j \\ \dfrac{ds}{d\tau} = \xi\cdot y+\eta\cdot x+\zeta , & \dfrac{dt}{d\tau} = \xi\cdot x-\eta\cdot y , \end{cases}$$

with initial conditions $x_j = y_j = s = t = 0$. Solving these equations and setting $\tau = 1$, we find that the point $e^{\lambda(\xi,\eta,\zeta)} \cdot 0 \in M$ has coordinates

$$(*) \qquad \begin{cases} z_j = \dfrac{1}{2}(\xi_j - i\eta_j) \\ w = \zeta + \dfrac{i}{4}(\xi^2 + \eta^2) . \end{cases}$$

This shows that <u>the map</u> $u \mapsto e^{\lambda(u)}.0$ <u>is a diffeomorphism from</u> G <u>onto</u> M .

The intertwining operator $W : C^{\infty}(M) \to C^{\infty}(\mathbf{G})$ defined by

$$Wf(u) = f(e^{\lambda(u)}.0)$$

can be expressed in these coordinates as

$$Wf(\xi,\eta,\zeta) = f(z_1,\ldots,z_n,w) \ ,$$

where $z_1,\ldots,z_n,w$ are given by (*)

Comments and references for Chapter II

§ 1.1 The inspiration for this section was the work of Folland-Stein [1] and Rothschild-Stein [1]. We have tried here to recast these constructions into a more geometric form, emphasizing the analogies with the case of a homogeneous space. For further differential-geometric aspects of this situation, cf. Tanaka [1], [2].

§ 1.2 The notion of "partial homomorphism" was introduced by Rothschild-Stein [1] for the case of the free nilpotent Lie algebras. For information about free Lie algebras and P.Hall bases, cf. Bourbaki [1].

§ 1.3 The "Lifting Theorem" was first stated and proved by Rothschild-Stein [1], from a different point of view and by somewhat different methods (the problem of "adding variables" to make vector fields "free up to step r.") The present treatment is taken from Goodman [7].

§ 2.1-2.2 The basic formula for the differential of the exponential mapping can be found in Helgason [1]. The formal inverse to this formula, involving the "Bernoulli operator," also appears in Berezin [1], Goodman [3], and Conze [1].

§ 2.3-2.4 One can reverse the order of construction here, and obtain the _a priori existence_ of the vector fields $\{Z_i\}$ by the implicit function theorem. This was pointed out to the author by P. Cartier. The formal series solution then serves to calculate the Taylor expansion of the coefficients of the $\{Z_i\}$, and the lifting Λ is constructed from the Z_i as in the text, cf. Goodman [7]. This procedure has the advantage of being applicable in any category where the implicit function theorem applies, e.g. C^∞, real analytic, or complex analytic.

§ 3.1-3.2 Theorem 3.2 is implicit in the paper of Rothschild-Stein [1]. The presentation here is taken from Goodman [7].

§ 4 The results and proofs here are taken from Kohn [3] and Bloom-Graham [1], who also obtain similar results for surfaces of codimension $k > 1$ which are in "general position". For the classification of the invariants of real hypersurfaces under holomorphic coordinate transformations, cf. Chern-Moser [1].

Chapter III

Singular integrals on spaces of homogeneous type

In this chapter we shall construct a general theory of "singular integral operators" on a class of locally compact spaces of "homogeneous type." Such a space X will have a "distance function" ρ and a measure μ related by the condition that the balls of radius R have measure of the order R^Q, where Q is a positive number (the "homogeneous dimension" of the space). The distance function ρ is required to satisfy a certain Lipschitz-continuity condition, which serves as a replacement for the triangle inequality.

The operators we shall study are of the form

$$Tf(x) = PV \int_X K(x,y)f(y)\, d\mu(y) ,$$

where K is a kernel which is singular along the diagonal (PV = principal value). The major result of this chapter is that when K satisfies certain homogeneity conditions, mean-value conditions, and smoothness conditions, then T defines a bounded operator on $L_2(X, d\mu)$. This is a generalization of the Calderon-Zygmund theory of singular integral operators, which applies both to nilpotent Lie groups and to the group germs studied in Chapter II. To make this connection, we begin, as in Chapter I, with some basic differential and integral calculus on graded vector spaces with dilations and homogeneous norms.

§ 1. Analysis on vector spaces with dilations

1.1. Homogeneous functions and distributions. Let V be a real, finite-dimensional vector space, with a direct-sum decomposition

$$V = \sum_{n=1}^{r} \oplus V_n .$$

Define the dilations $\{\delta_t : t > o\}$ on V as in Ch I, § 1.1, and denote Lebesgue measure on V by dx. We then have the integration formula

$$t^Q \int_V f(\delta_t x)\, dx = \int_V f(x)\, dx \quad ,$$

where $Q = \Sigma\, n \dim (V_n)$. We shall refer to the integer Q as the <u>homogeneous dimension</u> of V. (If $V = V_1$, then $Q = \dim V$).

A function f on V will be called <u>homogeneous of degree</u> μ $(\mu \in \mathbb{C})$, if

$$f \circ \delta_t = t^\mu f \qquad (t > o) .$$

A distribution g on V will be called <u>homogeneous of degree</u> μ $(\mu \in \mathbb{C})$ if

$$t^Q \langle f \circ \delta_t , g\rangle = t^{-\mu} \langle f,g\rangle$$

for all $f \in C_c^\infty (V)$. For example, the integration formula above shows that the measure dx on V defines a distribution which is homogeneous of degree 0. If g is a locally integrable function which is homogeneous of degree μ, then the distribution $g(x)\, dx$ is homogeneous of degree μ.

Fix a basis $\{x_i : 1 \le i \le d\}$ for V such that $x_i \in V_{n_i}$, and let $\{\xi_i\}$ be the dual basis for V^*. Write $D_i = \partial/\partial\xi_i$, and define

$$\|\nabla f\|_\infty = \sup \{|D_i f(x)| : |x| = 1,\ 1 \le i \le d\} .$$

Here $|x|$ is any homogeneous norm on V. Then we have the following version of the mean-value theorem for homogeneous functions.

<u>Lemma</u>. Suppose f is a C^1 function on $V \sim \{o\}$ which is homogeneous of degree μ. There exists a constant M, independent of f, such that

$$|f(x) - f(y)| \le M\, \|\nabla f\|_\infty\, |x|^{\lambda-1}\, |x-y| \, ,$$

when $|x| \ge M\, |x-y|$ (Here $\lambda = \mathrm{Re}\, \mu$).

<u>Proof</u>. If g is a function on $V \sim \{o\}$ which is homogeneous of degree μ,

then

$$|g(x)| \leq \|g\|_\infty \, |x|^\lambda ,$$

where $\|g\|_\infty = \sup \{|g(x)| : |x| = 1\}$, and $\lambda = \operatorname{Re} \mu$. Let us apply this to the estimate for $f(x) - f(y)$ obtained by integrating df along the path $\gamma(t) = x+t\,(y-x)$, $0 \leq t \leq 1$. We observe that $D_i f$ is homogeneous of degree $\mu - n_i$. Hence

$$|f(x) - f(y)| \leq \sup_{0 \leq t \leq 1} \Sigma |D_i f(\gamma(t))| \, |\xi_i(x-y)|$$

$$\leq M \, \|\nabla f\|_\infty \, \{ \sup_{0 \leq t \leq 1} \Sigma \, |\gamma(t)|^{\lambda - n_i} \, |x-y|^{n_i}\} ,$$

provided the path γ does not pass through 0 $(M = \max \|\xi_i\|_\infty)$.

To continue these estimates, we note that $|x+y| \leq K\,(|x| + |y|)$ and $|tx| \leq K|x|$ when $0 \leq t \leq 1$, where $K \geq 1$. Hence

$$|x| \leq K\,(|\gamma(t)| + |x-y|)$$

$$|\gamma(t)| \leq K^2(|x| + |x-y|) ,$$

so that if $|x| \geq 2K\,|x-y|$, then

$$(2K)^{-1} \, |x| \leq |\gamma(t)| \leq 2K^2 |x| .$$

In particular, the path γ does not pass through 0 , and

$$|\gamma(t)|^{\lambda - n_i} \, |x-y|^{n_i} \leq C_0 \, |x|^{\lambda - n_i} \, |x-y|^{n_i}$$

$$\leq C_1 \, |x|^\lambda \, |x-y| ,$$

provided $|x| \geq 2K\,|x-y|$, since $n_i \geq 1$. Here C_0 and C_1 are constants depending only on K , λ , and n_i . This proves the Lemma.

Corollary. Suppose the homogeneous norm $|x|$ is C^1 on $V \sim \{o\}$. Then it is Lipschitz continuous:

$$||x| - |y|| \leq C\,|x-y| .$$

1.2 Integral formulas. Suppose $|x|$ is a homogeneous norm on V. Then we have the following integration formula:

$$\int_{A \leq |x| \leq B} |x|^{-Q}\, dx = C \log (B/A) ,$$

where C is a constant independent of A,B, and Q is the homogeneous dimension of V.

To prove this formula, set

$$f(t) = \int_{1 \leq |x| \leq t} |x|^{-Q}\, dx \qquad (t \geq 1) .$$

Because the measure $|x|^{-Q}\, dx$ is invariant under dilations, we find that for $s,t \geq 1$,

$$f(st) = f(s) + f(t) .$$

In particular, this implies that $f(1) = 0$. If we define $f(t) = -f(1/t)$, $0 < t < 1$, then f is a continuous homomorphism from the multiplicative group $\mathbb{R}_+$ to the additive group $\mathbb{R}$. Hence $f(t) = C \log t$, as asserted.

More generally, suppose that ω is any continuous function on $V \sim \{o\}$ which is homogeneous of degree zero. Then there exists a constant $m(\omega)$ such that

$$\int_V f(|x|)\, \omega(x)\, |x|^{-Q}\, dx = m(\omega) \int_{\mathbb{R}_+} f(t)\, t^{-1}\, dt$$

for any $f \in L_1(\mathbb{R}_+ ; t^{-1}\, dt)$. ($\mathbb{R}_+ = (0,\infty)$). (The case f = characteristic function of $[A,B]$ is proved as above, and the general case follows by dominated convergence.)

We shall call $m(\omega)$ the mean-value of ω. The map $\omega \to m(\omega)$ is a continuous linear functional on the space of continuous functions homogeneous of degree 0, relative to the norm $\|\omega\|_\infty$. Indeed, we have, for $R > 1$,

$$m(\omega) \log R = \int_{1 \leq |x| \leq R} \omega(x)\, |x|^{-Q}\, dx ,$$

so that

$$|m(\omega)| \log R \leq \|\omega\|_\infty \int_{1\leq|x|\leq R} |x|^{-Q}\, dx$$

and hence

$$|m(\omega)| \leq \|\omega\|_\infty .$$

(We normalize the measure dx on V by the condition $m(1) = 1$.)

§ 2. Spaces of homogeneous type

2.1 Distance functions Let X be a locally compact Hausdorff space. Suppose

$$\rho : X \times X \to [0,\infty) .$$

We shall call ρ a distance function on X provided the following conditions are satisfied:

1) $\rho(x,y) \neq 0$ if $x \neq y$, and $\rho(x,x) = 0$
2) $\rho(x,y) = \rho(y,x)$
3) The sets $\{y : \rho(x,y) \leq r\}$, for $r > 0$, are compact and are a basis for the neighborhoods of x
4) There exists constants $C > 0$ and $0 < a \leq 1$ such that $|\rho(x,y) - \rho(x,z)| \leq C\, \rho(y,z)^a\, [\rho(x,y) + \rho(y,z)]^{1-a}$.

We shall call $\rho(x,y)$ the distance between x and y . The number a will be called the exponent of ρ . Note that a is required to be strictly positive, so that $1 - a < 1$.

Examples 1. Let V be a graded real vector space as in § 1.1 , with dilations δ_t . Let $|\cdot|$ be a smooth, symmetric homogeneous norm on V , and set $\rho(x,y) = |x-y|$. Then by Corollary 1.1 ,

$$|\rho(x,y) - \rho(x,z)| \le C\ |x-y - x+z| = |y-z| .$$

Thus 4) is satisfied with $a = 1$. The other properties are obviously verified.

2. More generally, let X be any simply connected nilpotent Lie group whose Lie algebra V admits a gradation (as a Lie algebra). Take a smooth, symmetric homogeneous norm $|\cdot|$ on V as in example 1, and define

$$\theta(x,y) = \log\ (x^{-1}y) \ , \quad \rho(x,y) = |\theta(x,y)| \ ,$$

where log is the inverse to the exponential map (If we define the group structure on V using the Campbell-Hausdorff formula, then log is the identity map). It is clear that conditions 1) - 3) are satisfied. To establish estimate 4) in this case, set $u = x^{-1}y$, $v = x^{-1}z$, $w = y^{-1}z$, and identify X with V as noted. Then $\theta(x,y) = u$, $\theta(x,z) = v$, $\theta(y,z) = w$. But $v = uw$, so by Corollary I.3.1 we have

$$|v-u-w| \le C\ (|u|^{a}|w|^{1-a} + |u|^{1-a}|w|^{a}) \ ,$$

where $a = 1/r$, r = length of the gradation on V . Further, for any norm function one has

$$|u-v| \le K(|u-v+w| + |w|) \ .$$

By considering the cases $|u| \ge |w|$ and $|u| \le |w|$ separately, we see that this gives the estimate

$$|u-v| \le C\ \{|u|^{1-a}|w|^{a} + |w|\} \ .$$

(Note that we may assume $a \le 1/2$, since the case $a = 1$ means X is commutative and $|u| = \|u\|$.) By Corollary 1.1, this gives condition 4) . The other properties are obvious.

3. Let λ be a partial homomorphism from a graded nilpotent Lie algebra V

into the Lie algebra of vector fields on a manifold M, as in Chapter II. Assume that for $x \in M$, the map $u \to \lambda(u)_x$ is a linear isomorphism from V onto the tangent space at x. Let the exponential map from V to M be defined as in § II.3. The analogue of the map $x,y \to \log(x^{-1}y)$ considered in the previous example is the map

$$\theta : M_0 \times M_0 \to V$$

defined implicitly by the identity

$$\exp_x(\theta(x,y)) = y .$$

Here $M_0 \subset M$ is a sufficiently small neighborhood of a given point $x_0 \in M$, as in § II.3.1.

Set $X = M_0$, $\rho(x,y) = |\theta(x,y)|$. Then ρ obviously satisfies 1) - 3), since the maps $\exp_x$, $x \in M_0$, are all diffeomorphisms. The verification of condition 4) is made by exactly the same argument as in the preceding example, but this time using Theorem II.3.2 to compare the group germ generated by $\lambda\cdot$ with the additive group of V.

4. Suppose X is <u>any</u> simply-connected nilpotent Lie group. Let F be a filtration on the Lie algebra V of X, and put a graded vector space structure on V by choosing complementary subspaces to the filtration, as in Chapter I. Let $\rho(x,y) = |\log(x^{-1}y)|$, as in example 2. Then ρ satisfies 1) - 3), and it satisfies 4) in every subset where $\rho(x,y) \geq \varepsilon \geq 0$, with a constant $C = C(\varepsilon)$. Indeed, using the same notation as in example 2 above, we have $|u| = \rho(x,y) \geq \varepsilon$. Hence by Theorem I.3.1 and Remark I.3.1, we get the same estimate for $|v-u-w|$ as in the graded case.

<u>Remarks</u> 1. Suppose ρ is a distance function on X. By the arithmetic mean-geometric mean inequality we have

$$|\rho(x,y) - \rho(x,z)| \leq K[\rho(x,y) + \rho(y,z)] .$$

In particular,

$$\rho(x,z) \leq 2K\,[\rho(x,y) + \rho(y,z)] ,$$

so that ρ satisfies a weak form of the triangle inequality.

2. Condition 4) on ρ will be used in situations where $\rho(y,z)$ is small compared to $\rho(x,y)$, i.e. y is close to z but far from x . In this case 4) implies that z is also far from x and that the difference $|\rho(x,y) - \rho(x,z)|$ is of smaller order of magnitude than the distance $\rho(x,y)$ (recall that we require $a > o$) .

3. In all the examples 2) - 4) above the number $r = 1/a$ is a positive integer equal to the length of the gradation on V . The longer the gradation, the weaker estimate 4) becomes, in terms of the comparison of distances described in the previous remark.

2.2 <u>Homogeneous measures</u> Let X, ρ be as in § 2.1. Suppose μ is a positive Radon measure on X . We shall say that μ is of <u>homogeneous type</u>, relative to the distance function ρ , if there exist constants $C, Q > 0$ such that

$$(*) \qquad \int_{A\leq\rho(x,y)\leq B} \rho(x,y)^{-Q}\, d\mu(y) \leq C \log (B/A)$$

for all $x \in X$, and all $0 < A \leq B$.

<u>Lemma</u> Assume μ satisfies (*) . Let $\alpha > 0$, $0 < A < B$. Then

(i) $\mu(\{y : \rho(x,y) \leq B\}) \leq CB^Q$

(ii) $\displaystyle\int_{\rho(x,y)\leq B} \rho(x,y)^{-Q+\alpha}\, d\mu(y) \leq (C/\alpha)\, B^\alpha$

(iii) $\displaystyle\int_{A\leq\rho(x,y)\leq B} \rho(x,y)^{-Q-\alpha}\, d\mu(y) \leq (C/\alpha)\, (A^{-\alpha} - B^{-\alpha})$.

Here the constant C is independent of α, A, B.

Proof. We estimate the integrals by the classical method of decomposition of the region of integration into concentric shells. Take $R < 1$, and for $k \in \mathbb{N}$ define.

$$S_k = \{y : BR^{k+1} \leq \rho(x,y) \leq BR^k\} .$$

Then by (*),

$$\int_{S_k} \rho(x,y)^{-Q+\alpha} \, d\mu(y) \leq CB^{\alpha} R^{k\alpha} \log(1/R) .$$

$$\int_{S_k} \rho(x,y)^{-Q-\alpha} \, d\mu(y) \leq CB^{-\alpha} R^{-(k+1)\alpha} \log(1/R) .$$

Now the spheres $\{x : \rho(x,y) = r\}$ have μ-measure zero, by (*), for any $r > o$ and any $y \in X$. In particular, $\mu(\{x\}) = 0$ for every x. Hence taking $\alpha = Q$ in the estimates above, we have

$$\mu(\{y : \rho(x,y) \leq B\}) \leq \sum_{k=o}^{\infty} \mu(S_k)$$

$$\leq C_1 B^Q$$

where C_1 is independent of B. The same argument gives (ii).

To prove (iii), we choose n so that $BR^{n+1} \leq A \leq BR^n$. Then the estimates above give

$$\int_{A \leq \rho(x,y) \leq B} \rho(x,y)^{-Q-\alpha} \, d\mu(y) \leq CB^{-\alpha} \log(1/R) \sum_{k=o}^{n} R^{-(k+1)\alpha}$$

$$\leq \frac{C \log(1/R)}{1 - R^{\alpha}} \left[(RA)^{-\alpha} - B^{-\alpha}\right]$$

$$\leq (C/\alpha) \left[(RA)^{-\alpha} - B^{-\alpha}\right] .$$

Letting $R \to 1$, we obtain (iii), Q.E.D.

<u>Remark</u> If we let $\alpha \to o$ in (iii), we regain estimate (*).

We shall call the triple (X,ρ,μ) satisfying the above conditions a <u>space of homogeneous type</u>, and we shall refer to the number Q as the <u>homogeneous dimension</u> of (X,ρ,μ). (In all the applications the estimate (*) will be an asymptotic equality when either $A \to 0$ or $B \to \infty$. In this case it is clear from the Lemma that Q is uniquely determined by (*).) Note that by part (i) of the Lemma a ball of radius R has measure $\leq CR^Q$, so this use of the term dimension has a geometric justification. When the choice of μ is clear from the context, we will write $d\mu(y) = dy$.

<u>Examples</u> 1. Let X be a simply-connected nilpotent Lie group, $\rho(x,y) = |\log(x^{-1}y)|$, as in § 2.1, Examples 1, 2 or 4. Let μ be Haar measure on X. Identifying X with its Lie algebra V by the exponential map, we can take Lebesgue measure on V for μ. Then

$$\int_{A\leq\rho(x,y)\leq B} \rho(x,y)^{-Q}\, d\mu(y) = \int_{A\leq|v|\leq B} |v|^{-Q}\, dv ,$$

by translation-invariance. Hence (*) follows from the integral formula of § 1.2. Furthermore, the inequalities in the Lemma become equalities in this case, by the same integration formula. If the gradation on V is a Lie-algebra gradation, then (X,ρ,μ) is a space of homogeneous type, in the sense just defined. If V is only filtered, but not graded, as a Lie algebra, then (X,ρ,μ) is of homogeneous type "near infinity", in the sense that ρ only satisfies condition 4) in § 2.1 when (x,y) is away from the diagonal in $X \times X$.

2. Let X,θ be as in § 2.1, Example 3. Take for μ any measure on X which in local coordinates has smooth, strictly positive density relative to Lebesgue measure. For $x \in X$, the map $y \longmapsto \theta(x,y)$ is a diffeomorphism from X onto an open, relatively compact subset U of V. The image of $d\mu$ under this mapping is of the form $\varphi(x,u)du$, where du is Lebesgue measure on V

and φ is a bounded, positive, smooth function on $X \times U$. Hence we have the integral formula

$$\int_X f(|\theta(x,y)|)\, d\mu(y) = \int_U f(|u|)\, \varphi(x,u)du ,$$

valid for any Borel function f on $\mathbb{R}_+$. By the integral formulas of § 1.2 this implies that μ and $\rho(x,y) = |\theta(x,y)|$ satisfy condition (*), where Q is the homogeneous dimension of the graded nilpotent algebra V.

2.3 Lipschitz spaces Let (X,ρ,μ) be a space of homogeneous type. A complex-valued function f on X satisfies a uniform Lipschitz condition of order β if there is a constant C such that

$$|f(x) - f(y)| \leq C\, \rho(x,y)^{\beta}$$

for all $x,y \in X$. In general, for any $\beta > 0$ we define

$$\|f\|_{\beta} = \sup_X |f(x)| + \sup_{x \neq y} \{\rho(x,y)^{-\beta}\, |f(x) - f(y)|\} .$$

Set $\Lambda(\beta) = \{f : \|f\|_{\beta} < \infty\}$. It is a standard exercise to verify that $\Lambda(\beta)$ is a Banach algebra relative to the norm $\|f\|_{\beta}$. Let

$$\Lambda_c(\beta) = \{f \in \Lambda(\beta) : \text{Supp}(f) \text{ is compact}\} .$$

Then by axiom 3) for the distance function, we see that $\Lambda_c(\beta) \subseteq C_o(X)$, the space of continuous functions vanishing at infinity on X.

Lemma Let a be the exponent of (X,ρ,μ). Then $\Lambda_c(\beta)$ is dense in $C_o(X)$ when $o < \beta \leq a$.

Proof. Let $\varphi \in C_c^{\infty}(\mathbb{R})$ and set $f(y) = \varphi(\rho(x,y))$. By axiom 3) for ρ and the local compactness of X, the function f will be compactly supported.

Since φ is smooth, we obtain from axiom 4) the estimate

$$|f(y) - f(z)| \leq C[\rho(y,z)^a \rho(x,y)^{1-a} + \rho(y,z)] .$$

This shows that $f \in \Lambda(a)$. Hence the functions in $\Lambda_c(\beta)$ separate the points of X if $0 < \beta \leq a$ (Note that $\Lambda_c(a) \subseteq \Lambda_c(\beta)$). As remarked above, one easily verifies that

$$\|fg\|_\beta \leq \|f\|_\beta \|g\|_\beta ,$$

so that $\Lambda_c(\beta)$ is a self-adjoint subalgebra of $C_0(X)$. The Stone-Weierstrass theorem thus gives the asserted density statement.

Corollary If $0 < \beta \leq a$, the space $\Lambda_c(\beta)$ is dense in $L_p(X,d\mu)$ for $1 \leq p < \infty$.

§ 3. Singular integral operators

3.1 Singular kernels. Let (X,ρ,μ) be a space of homogeneous type, with homogeneous dimension Q and exponent a. Let $D = \{(x,x) : x \in X\}$ be the diagonal in $X \times X$, and assume that

$$K : X \times X \sim D \to \mathbb{C}$$

is a continuous function. Define

$$K^*(x,y) = K(y,x)^c$$

(c = complex conjugate)

Definition. K is a singular kernel on X if there exist constants $C, M, R > 1$ such that the following estimates are satisfied by $F = K$ and $F = K^*$:

(I) $$|F(x,y)| \leq C\,\rho(x,y)^{-Q}$$

(II) If $$I_F(A,x) = \int_{A \leq \rho(x,y) \leq AR} F(x,y)\,dy ,$$

then for all $x \in X$,

$$\begin{cases} |I_F(A,x)| \leq C\,A , & \text{if } A \leq R \\ I_F(A,x) = 0 , & \text{if } A \geq R \end{cases}$$

(III) If $\rho(y,z) \geq M\,\rho(x,z)$, then

$$|F(y,z) - F(y,x)| \leq C \left[\frac{\rho(x,z)}{\rho(y,z)}\right]^a \rho(y,z)^{-Q} .$$

We shall call (I) the homogeneity condition, (II) the mean-value condition, and (III) the smoothness condition. Before giving examples of singular kernels, which will explain the terminology, we recast these estimates from the point of view of functional analysis. Namely, we fix $M, R > 1$ and define the following norms:

$$\|K\|_I = \sup\ \rho(x,y)^Q\,|K(x,y)|$$

$$\|K\|_{II} = \sup\ A^{-1}\ |I_K(A,x)|$$

$$\|K\|_{III} = \sup \left[\frac{\rho(y,z)}{\rho(x,z)}\right]^a \rho(y,z)^Q\,|K(y,z) - K(y,x)| ,$$

the suprema being taken over x,y,z,A such that $x \neq y$, $x \neq z$, $0 < A \leq R$, and in the case of $\|K\|_{III}$, $\rho(y,z) \geq M\,\rho(x,z)$.

Define the space

$$\mathcal{K}_{M,R}(X,\rho,\mu) = \{K : K \text{ and } K^* \text{ satisfy I,II,III for some } C\} .$$

We norm this space by setting $\|K\|_{M,R} = \|K\| + \|K^*\|$, where $\|K\| = \|K\|_I + \|K\|_{II} + \|K\|_{III}$. Then it is straightforward to verify that $\mathcal{K}_{M,R}$ is a Banach space in this norm. For example, condition III for K and K^* implies that K is continuous on $X \times X \sim D$. Also the norm $\|K\|_I$ dominates the sup norm on any compact subset of $X \times X \sim D$, and hence also dominates the functionals $K \longmapsto I_K(A,x)$. Thus the continuity of K and the vanishing mean-value condition are preserved under limits in the norm $\|K\|_{M,R}$.

<u>Examples</u> Consider two classes of examples from § 2.2, namely the graded nilpotent group case (example 1), or the group germ case (example 2). In both cases we have a graded vector space V of homogeneous dimension Q and a mapping

$$\theta : X \times X \to V ,$$

such that

$$\rho(x,y) = |\theta(x,y)| ,$$

where $|\cdot|$ is a smooth, symmetric homogeneous norm on V. Let Q be the homogeneous dimension of V. We obtain a class of singular kernels as follows:

Let λ be a complex number, and define $K_\lambda(V)$ to be the space of C^1 functions on $V \sim \{o\}$ which are homogeneous of degree λ, relative to the dilations on V. Give this space the norm

$$\|k\|_\infty + \|\nabla k\|_\infty ,$$

where $\|k\|_\infty = \sup \{|k(x)| : |x| = 1\}$, as in § 1.1. Then $K_\lambda(V)$ is a Banach space.

<u>Lemma</u> Suppose $\lambda = -Q + is$, with $s \neq o$ a real number. Set $r = \exp(2\pi/|s|)$.

Then there exists an $M > o$ and an integer $n \geq 1$ so that map $k \longmapsto k \circ \theta$ is continuous from $K_\lambda(V)$ into $K_{M,R}(X,\rho,\mu)$, with $R = r^n$. When $\lambda = -Q$, this map is continuous from the subspace $K^o_{-Q}(V)$ of functions with mean-value zero into $K_{M,R}(X,\rho,\mu)$ for all sufficiently large R.

Proof Let $k \in K_\lambda(V)$, and set $K(x,y) = k(\theta(x,y))$. Then

$$|K(x,y)| \le \|k\|_\infty \rho(x,y)^{-Q} ,$$

and hence $\|K\|_I \le \|k\|_\infty$.

To check condition (II), we observe that the function $|x|^{-\lambda} k(x) = \omega(x)$ is homogeneous of degree zero. Hence by the integral formula of § 1.2, if $R = \exp(2\pi n/|s|)$, then

$$\int_{A\le|x|\le AR} k(x)\, dx = m(\omega) \int_A^{AR} t^{-1+is}\, dt$$

$$= m(\omega)\left[A^{is} (R^{is} - 1)/(is)\right] .$$

By the choice of R, $R^{is} = 1$, so this vanishes. In case $s = o$, then $m(\omega) = o$ by assumption.

In the nilpotent group case the translation invariance of μ gives the formula

$$I_K(A,x) = \int_{A\le|u|\le AR} k(u)\, du = 0 .$$

In the group germ case, we can write

$$I_K(A,x) = \int_{A\le|u|\le AR} k(u) \left[\varphi(x,u) - \varphi(x,o)\right] du .$$

By the compact support of φ, this vanishes for $A \ge R = r^n$ if n is sufficiently large. When $A < R$ we can estimate $\varphi(x,u) - \varphi(x,o)$ and get the bound

$$|I_K(A,x)| \le C \|k\|_\infty \int_{A\le|u|\le AR} |u|^{-Q+1}\, du$$

$$\le C A \|k\|_\infty .$$

This gives the estimate for $\|K\|_{II}$.

By Lemma 1.1, we have the estimate

$$|K(y,z) - K(y,x)| \leq C \|\nabla k\|_\infty |\theta(y,z) - \theta(y,x)| \rho(y,z)^{-Q-1} ,$$

when $\rho(y,z) \geq C |\theta(y,z) - \theta(y,x)|$. In order to obtain estimate (III) for K from this inequality, we recall that

$$|\theta(y,z) - \theta(y,x)| \leq C \rho(x,z)^a [\rho(x,z) + \rho(y,z)]^{1-a}$$

(cf § 2.1, Examples) . This estimate is valid for any x,y,z in the group germ or graded nilpotent group case. (It is valid for $\rho(x,y) \geq \varepsilon > 0$ in the case of an arbitrary nilpotent group.) But now if $\rho(y,z) \geq M \rho(x,z)$, then

$$|\theta(y,z) - \theta(y,x)| \leq C M^{-1} [M + 1]^{1-a} \rho(y,z) .$$

Since $1-a < 1$, we can choose M so large that $M^{-1} [M + 1]^{1-a} < C^{-2}$. Then when $\rho(y,z) \geq M \rho(x,z)$, the above estimate for $|K(y,z) - K(y,x)|$ applies, and we get

$$\|K\|_{III} \leq C \|\nabla k\|_\infty .$$

(Here a is the exponent of ρ , in III , and the choice of M is independent of k.) Since $\theta(y,x) = -\theta(x,y)$, the same estimates hold for K^* . This completes the proof of the Lemma.

3.2 Operators defined by singular kernels Let $K \in K_{M,R}(X,\rho,\mu)$ be a singular kernel. In order to define an integral operator with kernel K , we must use a limiting process. Given $m < n \in \mathbb{Z}$, we define the truncated operator

$$K_{m,n} f(x) = \int_{R^m \leq \rho(x,y) \leq R^n} K(x,y) f(y) \, dy .$$

The integral converges absolutely for any bounded, measurable function f , and

$$\|K_{m,n} f\|_\infty \leq C_{m,n} \|f\|_\infty \|K\|_I .$$

(Here $\|f\|_\infty = \sup |f(x)|$, and $\|K\|_I$ was defined in the previous section.) This estimate, however, is of little use, since

$$C_{m,n} = \int_{R^m \leq \rho(x,y) \leq R^n} \rho(x,y)^{-Q} \, dy ,$$

and this is of the order $\log (R^{n-m})$. Letting $m \to -\infty$ or $n \to +\infty$ gives no information. The next result is a slight improvement over this crude estimate. (The space $\Lambda(\beta)$ was defined in § 2.3):

<u>Theorem</u> If $o < \beta \leq a$ and $f \in \Lambda_c(\beta)$, then

$$Kf = \lim_{\substack{m \to -\infty \\ n \to +\infty}} K_{m,n} f$$

exists in L_p for $1 < p \leq \infty$, and

$$\|Kf\|_{L_p} \leq C_p \, \mu(\mathrm{Supp}(f)) \, \|K\| \, \|f\|_\beta ,$$

where C_p is a constant independent of K and f .

<u>Proof</u> We shall write

$$K_m = K_{m,1} \quad , \quad m \leq 0$$

$$K^n = K_{1,n} \quad , \quad n \geq 1 .$$

Then $K_{m,n} = K_m + K^n$, and we shall investigate each summand separately.

Let $\varphi_m = K_m 1$, where 1 is the function identically 1 . By condition (II) on K we see that

$$\|\varphi_m\|_\infty \leq R \, \|K\|_{II} ,$$

and

$$\lim_{m \to -\infty} \varphi_m(x) = \varphi(x)$$

exists uniformly in x, with $\|\varphi\|_\infty \leq R\,\|K\|_{II}$.

We may write

$$K_m f(x) = \int_{R^m \leq \rho(x,y) \leq R} K(x,y)\,|f(y) - f(x)|\,dy + \varphi_m(x) f(x)\,.$$

Hence if $m < n \leq 0$, we have

$$|K_m f(x) - K_n f(x)| \leq \|K\|_I\,\|f\|_\beta \int_{R^m \leq \rho(x,y) \leq R^n} \rho(x,y)^{-Q+\beta}\,dy$$

$$+ |I_K(R^{m-n},x)|\;\|f\|_\beta$$

$$\leq CR^{(m-n)a}(\|K\|_I + \|K\|_{II})\,\|f\|_\beta\,,$$

by Lemma 2.2. This shows that

$$\lim_{m\to-\infty} K_m f(x) = K_\infty f(x)$$

exists uniformly in x. If f is compactly supported, so is $K_m f$, uniformly in $m \leq 0$. Hence this limit exists in all L^p norms, $1 \leq p \leq \infty$.

To estimate $K^n f$, we use the Schwarz inequality to obtain, if $0 \leq m < n$,

$$|K^m f(x) - K^n f(x)| \leq \|K\|_I\,\|f\|_{L_2}\,\Big\{\int_{R^m \leq \rho(x,y) \leq R^n} \rho(x,y)^{-2Q}\,dy\Big\}^{1/2}$$

$$\leq C\,R^{-Qm}\,\|K\|_I\,\|f\|_{L_2}\,,$$

by Lemma 2.2. This shows that

$$\lim_{n\to\infty} K^n f(x) = K^\infty f(x)$$

exists uniformly in x.

To show that this limit exists in L_p, $1 < p < \infty$, we observe that the kernel $K(x,y)$, truncated in the region $\rho(x,y) < R$, is in L_p relative to x, for fixed y, and vice-versa. Let $g \in L_q$, where $1/p + 1/q = 1$. Then by Hölders inequality,

$$\left| \int_{\rho(x,y)\geq R} K(x,y)\, g(x)\, dx \right| \leq C_p \, \|K^*\|_I \, \|g\|_{L_q} ,$$

where

$$C_p = \int_{\rho(x,y)\geq R} \rho(x,y)^{-pQ} \, dx < \infty$$

by Lemma 2.2, since $p > 1$. (Here we have used condition I for K^*.) Thus if f is compactly supported, then

$$< K^n f, g > = \int_{y\in \mathrm{Supp}(f)} \int_{R\leq \rho(x,y)\leq R^n} K(x,y) f(y) g(x) \, dx \, dy .$$

By the estimate above, we have

$$| < K^n f, g > | \leq C_p \, \mu(\mathrm{Supp}(f)) \, \|K^*\|_I \, \|f\|_\infty \, \|g\|_{L_q} ,$$

where the constant C_p is independent of K, f, g. This shows that

$$\|K^n f\|_{L_p} \leq C_p \, \mu(\mathrm{Supp}(f)) \, \|K\| \, \|f\|_\infty ,$$

and completes the proof of the theorem.

§ 4. Boundedness of singular integral operators

4.1 Almost orthogonal decompositions Let H be a complex Hilbert space, and denote by $B(H)$ the algebra of bounded operators on H. If $T \in B(H)$, then T^* will denote the operator adjoint to T:

$$(Tx, y) = (x, T^* y) , \qquad x, y \in H .$$

We denote by $\|T\|$ the operator norm of T, and we recall the fundamental relations between the norm and the $*$-algebra structure on $B(H)$:

$$\|T\| = \|T^*T\|^{1/2} = \lim_{n\to\infty} \|(T^*T)^n\|^{1/2n} .$$

(The first equality is immediate from the definition of T^* and the Schwarz inequality. The second equality follows from the self-adjointness of T^*T and the spectral theorem.)

Suppose now that $\{T_k\}_{k\in\mathbb{Z}}$ is a family of operators on H . We will say that the $\{T_k\}$ are <u>almost orthogonal</u> if there exists a function $\varphi \in \ell_1(\mathbb{Z})$ such that $\varphi \geq 0$ and

$$(*) \qquad \|T_j^* T_k\| + \|T_j T_k^*\| \leq \varphi(j-k)^2$$

for all $j,k \in \mathbb{Z}$. (Set $\|\varphi\|_{\ell_1} = \sum_{k\in\mathbb{Z}} \varphi(k)$) .

For example, if $\{P_k\}$ is a family of mutually orthogonal projections on H , then $P_jP_k^* = P_j^*P_k = 0$ when $j \neq k$, and $\|P_jP_j^*\| = \|P_j\| = 1$. In this case estimate (*) is satisfied with $\varphi = 2\delta$ (δ = Kronecker delta). Furthermore, it is a basic fact in Hilbert space geometry that the series ΣP_j converges in $B(H)$, because of the orthogonality, even though each term has norm 1. The next theorem generalizes this geometric property to the case of almost orthogonal families of operators.

<u>Theorem</u> Suppose $\{T_k\}$ is a family of operators on H which satisfies (*). If $J \subset \mathbb{Z}$ is a finite subset, let

$$S_J = \sum_{k\in J} T_k .$$

Then

$$\|S_J\| \leq \|\varphi\|_{\ell_1} ,$$

and hence the partial sums S_J are uniformly bounded in $B(H)$.

Corollary: If $\lim_{|J|\to\infty} S_J x \equiv Sx$ exists for x in a dense subspace of H, then $S \in B(H)$ and $\|S\| \le \|\varphi\|_{\ell_1}$.

Proof of Theorem We shall estimate $\|(S_J^* S_J)^n\|$. This is bounded by the sum of terms

$$\|T_{i_1}^* T_{j_1} T_{i_2}^* T_{j_2} \cdots T_{i_n}^* T_{j_n}\| ,$$

where $i_k, j_k \in J$. Grouping the factors i_k, j_k together and using the submultiplicative property of the norm, we obtain from (*) the estimate

$$\varphi(i_1-j_1)^2 \varphi(i_2-j_2)^2 \cdots \varphi(i_n-j_n)^2$$

for such a term. Grouping the factors j_k, i_{k+1} together, we similarly obtain the estimate

$$\varphi(0) \varphi(j_1-i_2)^2 \cdots \varphi(j_{n-1}-i_n)^2 \varphi(0)$$

for the same term. (Note that $\|T_j\| = \|T_j^*\| = \|T_j^* T_j\|^{1/2} \le \varphi(0)$.) Taking the geometric mean of these two estimates, we see that the term is bounded by the product

$$\varphi(0) \varphi(i_1-j_1) \varphi(j_1-i_2) \cdots \varphi(j_{n-1}-i_n) \varphi(i_n-j_n) .$$

We may now sum this over all $i_k, j_k \in J$. This gives the estimate

$$\|(S_J^* S_J)^n\| \le \#(J)\, \varphi(0)\, \|\varphi\|_{\ell_1}^{2n-1} .$$

Taking the 2n-th root and letting $n \to \infty$, we obtain the stated inequality for $\|S_J\|$.

4.2 Decompositions of singular integrals Let (X,ρ,μ) be a space of homogeneous type, and let K be a singular kernel on X in the sense of § 3.1.

We want to use the Theorem of the preceding paragraph to establish that the operator defined by K is bounded on $L_2(X, d\mu)$. For this we shall decompose K into a sum of almost orthogonal operators.

Let the constants $M,R > 1$ be such that $K \in \mathcal{K}_{M,R}(X,\rho,\mu)$. For $j \in \mathbb{Z}$, define

$$K_j(x,y) = \begin{cases} K(x,y), \text{ if } R^j \leq \rho(x,y) \leq R^{j+1} \\ 0 \quad , \quad \text{otherwise} \end{cases},$$

and set

$$T_j f(x) = \int K_j(x,y)\, f(y)\, dy \ ,$$

for $f \in L_2(X, d\mu)$. Let $\|A\|$ denote the operator norm of an operator A on $L_2(X, d\mu)$. Then we have the following estimate which shows that the two families $\{T_j\}_{j\geq o}$ and $\{T_j\}_{j\leq o}$ are both almost orthogonal. (This separation of $j \geq o$ and $j \leq 0$ corresponds to the two singularities of K : the point at infinity and the diagonal.)

<u>Lemma</u> There is a constant C, independent of K, such that when j,ℓ have the same sign, then

$$\|T_j T_\ell^*\| + \|T_j^* T_\ell\| \leq CR^{-a|j-\ell|} \|K\|_{M,R}^2$$

(The norm $\|K\|_{M,R}$ was defined in § 3.1. Here a is the exponent of (X,ρ,μ).)

<u>Proof</u>. If A is an integral operator with kernel $A(x,y)$, then by the Schwarz inequality one has the pointwise estimate

$$|Af(x)|^2 \leq (\int|A(x,z)|dz) \int|A(x,y)||f(y)|^2\, dy \ .$$

Integrating this estimate, we find that

$$(*) \qquad \|A\| \leq \left[\sup_x \int|A(x,y)|dy\right]^{1/2} \left[\sup_y \int|A(x,y)|dx\right]^{1/2} .$$

Let us apply this estimate to T_j. By condition (I) on K, we have

$$\|T_j\| \le \|K\|_I \int_{R^j \le \rho(x,y) \le R^{j+1}} \rho(x,y)^{-Q}\, dy$$

$$\le (C \log R)\ \|K\|_I \quad ,$$

where C is independent of K. Thus the operators $\{T_j\}$ are all uniformly bounded, and it will suffice to prove the lemma when $|j-\ell| \ge N$, for any fixed N. (N will be chosen in the course of the proof and will depend on M,R, and the distance function ρ, but the choice will be independent of $K \in \mathcal{K}_{M,R}(X,\rho,\mu)$.)

The operator $T_j T_\ell^*$ is an integral operator with kernel

$$G_{j\ell}(x,y) = \int K_j(x,z)\ K_\ell(y,z)^c\ dz.$$

Since both K and K^* are assumed to satisfy (I), (II), (III) of § 3.1, the estimates we obtain for $\|T_j\ T_\ell^*\|$ will also apply to $\|T_j^*\ T_\ell\|$. Thus it is enough to estimate the right side of (*) when $A(x,y) = G_{j\ell}(x,y)$.

We shall establish the inequalities

$$(**)\quad \begin{cases} \sup_x \int |G_{j\ell}(x,y)|\,dy \le CR^{a(j-\ell)}\ \|K\|_{M,R}^2\ , & \text{if } \ell \ge j \\ \sup_y \int |G_{j\ell}(x,y)|\,dx \le CR^{a(\ell-j)}\ \|K\|_{M,R}^2\ , & \text{if } j \ge \ell\ , \end{cases}$$

provided that j,ℓ are of the same sign and $|j-\ell| \ge N$, where N is independent of K. Once this is done, we observe that

$$G_{j\ell}(x,y) = G_{\ell j}(y,x)^c\ .$$

Hence (**) implies that if $j\ell \ge 0$ and $|j-\ell| \ge N$, then

$$\begin{cases} \sup_x \int |G_{j\ell}(x,y)|\,dy \le CR^{-a|j-\ell|}\ \|K\|_{M,R}^2 \\ \sup_y \int |G_{j\ell}(x,y)|\,dx \le CR^{-a|j-\ell|}\ \|K\|_{M,R}^2\ . \end{cases}$$

Using this in (*), we will obtain the estimate of the lemma.

To prove (**) , we consider the cases $\ell \geq j$ and $j \geq \ell$ separately. Suppose first that $\ell \geq j$. Fix x , and set

$$E = \{(y,z) : R^j \leq \rho(x,z) \leq R^{j+1} , \quad R^\ell \leq \rho(y,z) \leq R^{\ell+1}\} .$$

All estimates take place on E .

We will make a three-term collapsing sum estimate of the left-hand side of (**) , corresponding roughly to the three conditions (I), (II), (III) of § 3.1 on the kernels K and K^* . Namely, we write

$$G_{j\ell}(x,y) = \int K_j(x,z) [K_\ell(y,z) - K(y,x)]^C dz$$

$$+ [K(y,x) - K_\ell(y,x)]^C \int K_j(x,z)\, dz$$

$$+ K_\ell(y,x)^C \int K_j(x,z)\, dz \quad .$$

Thus we have

$$\int |G_{j\ell}(x,y)| dy \leq \iint_E |K_j(x,z)||K(y,z) - K(y,x)| dy\, dz$$

$$+ \iint_E |K_j(x,z)||K(y,x) - K_\ell(y,x)| dy\, dz$$

$$+ |\int K_j(x,z)\, dz| \int |K_\ell(y,x)|\, dy$$

(Note that $K_\ell(y,z) = K(y,z)$ when $(y,z) \in E$.) Call the terms in this sum I_1, I_2, I_3, respectively. We shall estimate each term separately.

The term I_3 is the simplest to estimate. Indeed, by (I) ,

$$\int |K_\ell(y,x)|\, dy \leq \|K\|_I \quad (C \log R) ,$$

and by (II) ,

$$|\int K_j(x,z)\, dz| \leq \|K\|_{II} \cdot \begin{cases} CR^{aj} & , \text{ if } j \leq 0 \\ 0 & , \text{ if } j > 0 \end{cases} .$$

Thus if $j > 0$, then $I_1 = 0$, while if $j \leq 0$, then $\ell \leq 0$ also, since j and ℓ are assumed to have the same sign. Hence in any case we get the estimate

$$I_3 \leq CR^{a(j-\ell)} \, \|K\|_I \, \|K\|_{II} \, .$$

To estimate I_2, note that the integrand is zero when

$$R^{\ell} \leq \rho(x,y) \leq R^{\ell+1} \, .$$

Furthermore, when $(y,z) \in E$, then by axiom 4) on the distance function we have

$$\begin{aligned} |\rho(x,y) - \rho(z,y)| &\leq C\, \rho(x,z)^a \, [\rho(x,z) + \rho(y,z)]^{1-a} \\ &\leq CR^{1+aj+(1-a)\ell} \, [1+R^{j-\ell}] \\ &\leq CR^{\ell+a(j-\ell)} \, , \end{aligned}$$

since $j-\ell \leq 0$. Here C is a constant independent of j,ℓ. If we write

$$J_+ = \{y : (y,z) \in E \text{ for some } z, \text{ and } \rho(x,y) \geq R^{\ell+1}\} \, ,$$

then this last estimate shows that

$$J_+ \subset \{y : R^{\ell+1} \leq \rho(x,y) \leq R^{\ell+1} \, [1 + CR^{a(j-\ell)}]\} \, .$$

Similarly, if we write

$$J_- = \{y : (y,z) \in E \text{ for some } z, \text{ and } \rho(x,y) \leq R^{\ell}\} \, ,$$

then the same estimate shows that

$$J_- \subset \{y : R^{\ell}[1 - CR^{a(j-\ell)}] \leq \rho(x,y) \leq R^{\ell}\} \, .$$

We now choose N so that

$$CR^{-aN} \leq 1/2 \, ,$$

and we require that $|j-\ell| \geq N$. Then we can estimate I_2 as follows: Write

$$I_2 = \iint\limits_{\substack{(y,z)\in E \\ y\in J_+ \cup J_-}} |K_j(x,z)||K(y,x)| \, dy \, dz \, .$$

By condition (I) on the kernel K, we thus have

$$I_2 \le ||K||_I^2 \left\{ \int\limits_{R^j \le \rho(x,z) \le R^{j+1}} \rho(x,z)^{-Q} dz \right\} \left\{ \int\limits_{J_+ \cup J_-} \rho(x,y)^{-Q} dy \right\} .$$

By the homogeneity condition for the measure μ and the above estimates for J_+ and J_-, this gives the estimate

$$I_2 \le C \, \|K\|_I^2 \, \log \left(\frac{1+\varepsilon}{1-\varepsilon}\right) ,$$

where $\varepsilon = CR^{a(j-\ell)} \le 1/2$. But

$$\log \left(\frac{1+\varepsilon}{1-\varepsilon}\right) \le 3\varepsilon \quad ,$$

if $0 \le \varepsilon \le 1/2$, so this gives the desired estimate for I_2.

To estimate I_1, we must use the smoothness condition (III) on the kernel K. If $(y,z) \in E$, then

$$\rho(x,z) \le R^{j-\ell+1} \, \rho(y,z) \, .$$

We now impose the additional requirement on N that

$$R^{N-1} \ge M \quad ,$$

where M is the constant in condition (III). Then if $\ell-j \ge N$ and $(y,z) \in E$, we have

$$\begin{aligned} |K(y,z) - K(y,x)| &\le \|K\|_{III} \left(\frac{\rho(x,z)}{\rho(y,z)}\right)^a \rho(y,z)^{-Q} \\ &\le \|K\|_{III} \, R^{a(j-\ell)+a} \, \rho(y,z)^{-Q} \, . \end{aligned}$$

Thus

$$I_1 = \iint\limits_E |K_j(x,z)||K(y,z) - K(y,x)| \, dy \, dz$$

$$\leq \|K\|_{III} \, R^{a(j-\ell)} \iint_E |K_j(x,z)| \, \rho(y,z)^{-Q} \, dy \, dz$$

$$\leq C \, \|K\|_{III} \, \|K\|_I \, R^{a(j-\ell)} \quad .$$

This completes the estimates in the case $\ell \geq j$.

For the case $j \geq \ell$, we simply interchange the roles of x and y in the above argument, i.e. we write

$$G_{j\ell}(x,y) = \int [K_j(x,z) - K(x,y)] \, K_\ell(y,z)^c \, dz$$

$$+ [K(x,y) - K_j(x,y)] \int K_\ell(y,z)^c \, dz$$

$$+ K_j(x,y) \int K_\ell(y,z)^c \, dz \quad ,$$

and make the analogous estimates. This completes the proof of the lemma.

Remarks 1. Suppose the kernel K actually has mean value zero , in the sense that

$$\int_{A \leq \rho(x,y) \leq AR} K(x,y) \, dy = 0 \quad ,$$

for some $R > 1$ and all $A > 0$. Suppose the same also holds for K^* . Then $I_2 = I_3 = 0$ in the above argument, and

$$G_{j\ell}(x,y) = \int [K_j(x,z) - K(x,y)] \, K_\ell(y,z)^c \, dz$$

The geometric basis for the entire lemma is then that on the set E where the integration takes place, the ratio of the distances admits the bound

$$\frac{\rho(x,z)}{\rho(y,z)} \leq R^{j-\ell+1} \quad .$$

The smoothness condition (III) for K , on the other hand, is expressed in terms of this ratio. This explains the appearance of the quantity $j-\ell$ in the estimate for $\|T_j \, T_\ell^*\|$.

2. If we restricted our attention to kernels with mean value zero, in the

sense of the preceding remark, then it would be sufficient to know that the measure μ satisfied

$$\mu(\{y : \rho(x,y) \leq R\}) \leq CR^Q$$

for some constant C and all x,R. Indeed, in this case we would have the estimate

$$\int_{A\leq\rho(x,y)\leq B} \rho(x,y)^{-Q}\, d\mu(y) \leq C(B/A)^Q \quad ,$$

and the estimate of the integral I_1 in the proof above only needed this last estimate on the integral of $\rho(x,y)^{-Q}$. In order to perturb the mean-value-zero condition, however, we must replace $(B/A)^Q$ by $\text{Log}\,(B/A)$. The vanishing of this quantity as $A \to B$ then compensates for the non-zero mean-value.

3. For the case of kernels with mean-value zero, axiom 4) on the distance function is not needed in the proof of the lemma. This only entered into the estimation of I_2.

4.3 L_p Boundedeness $(1 < p < \infty)$ Let (X,ρ,μ) be a space of homogeneous type, and let $K_{M,R}(X,\rho,\mu)$ be the Banach space of singular kernels defined in § 3.1, with norm $\|K\|_{M,R}$. If $K \in K_{M,R}$ is a singular kernel, then by Theorem 3.2 we know that K defines a "principal-value" operator (also denoted by K) mapping the Lipschitz space $\Lambda_c(\beta)$ into $L_p(X,\mu)$, for $o < \beta \leq a$ and $1 < p < \infty$. We can now strenthen this result, as follows:

Theorem The operator K maps L_p continuously into L_p, for $1 < p < \infty$. More precisely, there exists a constant C_p, independent of K, such that

$$(*) \qquad \|Kf\|_{L_p} \leq C_p \|K\|_{M,R} \|f\|_{L_p} ,$$

for $1 < p < \infty$.

Proof By Lemma 4.2 and Corollary 4.1, we know that (*) holds when $p = 2$. Define the truncated kernel

$$K_\varepsilon(x,y) = \begin{cases} K(x,y), & \text{if } \varepsilon \leq \rho(x,y) \leq 1/\varepsilon , \\ 0 , & \text{otherwise} . \end{cases}$$

Then we claim that there exist constants M and C, independent of K and ε, such that

$$(**) \qquad \int_{\rho(x,z)\geq M\rho(y,z)} |K_\varepsilon(x,y) - K_\varepsilon(x,z)| \, dx \leq C \|K\| .$$

Indeed, if we calculate the same integral for the kernel K instead of K_ε, and use the smoothness condition (III) on K, we obtain the majorization

$$\|K\| \, \rho(y,z)^a \int_{\rho(x,z)\geq M\rho(y,z)} \rho(x,y)^{-Q-a} \, dx .$$

This in turn is majorized by $C \|K\|$ by part (iii) of Lemma 2.2. Thus we only need to estimate the error involved in replacing K_ε by K in (**). Note that because of the inequality $\rho(x,y) \leq \kappa[\rho(x,z) + \rho(z,y)]$, the integrand in (**) vanishes when $\rho(x,y)$ and $\rho(x,z)$ are outside an interval $(C^{-1}\varepsilon, C\varepsilon^{-1})$, where

$C \geq 1$ is a constant depending only on κ and M. From the homogeneity condition (I) on K, we can thus bound the error by

$$2 \|K\| \int_{A_\varepsilon \cup B_\varepsilon} \rho(x,y)^{-Q} \, dx ,$$

where

$$A_\varepsilon = \{x : C^{-1} \varepsilon \leq \rho(x,y) \leq \varepsilon\}$$

$$B_\varepsilon = \{x : \varepsilon^{-1} \leq \rho(x,y) \leq C\varepsilon^{-1}\} .$$

By the definition of a homogeneous measure, this integral is bounded by $C_1 \|K\|$, independently of ε. This proves (**).

With (*) for $p = 2$ and (**) established, the proof for $1 < p < 2$ follows from a "covering lemma", which proves that the operators K_ε are of weak type (1,1), uniformly in ε, and from the Marcinkiewicz interpolation theorem, which proves that the operators K_ε are bounded on L_p, $1 < p < 2$, with norm $\leq C \|K\|$. Since the same is true for the operators K_ε^*, the boundedness in the range $2 < p < \infty$ follows by duality (cf. Comments and references).

§ 5 Examples

5.1 Graded nilpotent groups Let G be a graded, simply-connected nilpotent Lie group, with dilations δ_t and smooth, symmetric homogeneous norm $|x|$. Let K_λ (resp. K^o_{-Q}) be the space of C^1 functions in $G \sim \{e\}$ which are homogeneous of degree $\lambda = -Q + is$, with Q = homogeneous dimension of G, and real $s \neq o$ (resp. homogeneous of degree $-Q$ with mean-value zero). From Lemma 3.1 and Theorem 4.3, we conclude that the kernel

$$K(x,y) = k(x^{-1}y)$$

defines a bounded operator K on $L_p(G)$, $1 < p < \infty$, when k is a function in $\mathcal{K}_\lambda$ or $\mathcal{K}^0_{-Q}$. Evidently K commutes with left translation by elements of G. Under dilations, K transforms by

$$K(\varphi\circ\delta_t) = t^{-is}\,(K\varphi)\circ\delta_t \;.$$

In particular, when $k \in \mathcal{K}^0_{-Q}$, then K commutes with dilations.

For the applications to group representations in Chapter IV, it will be important to know that for homogeneous kernels of degree $-Q$, the vanishing mean-value of the kernel is not only a sufficient, but also a <u>necessary</u> condition, for boundedness on $L_2(G)$ of a singular convolution operator. Since any such kernel k can be written as

$$k(x) = k_0(x) + c|x|^{-Q} \;,$$

where k_0 has mean-value zero, it suffices to consider the particular kernel $|x|^{-Q}$.

<u>Theorem</u> Let μ be any distribution on G such that $\mu = |x|^{-Q}$ on $G \sim \{e\}$. Then convolution by μ (defined on $C_c^\infty(G)$) does not extend to a bounded operator on $L_2(G)$.

<u>Proof</u> (Sketch) Since $|x|^{-Q}$ is not integrable at $x = e$, it must be "regularized" at e to give a distribution. One such regularization is evidently the distribution

$$(*) \qquad < T,\varphi > = \int_{|x|\le 1} \{\varphi(x) - \varphi(e)\}\,\frac{dx}{|x|^Q} + \int_{|x|\ge 1} \varphi(x)\,\frac{dx}{|x|^Q} \;.$$

If μ is any other distribution such that $\mu = |x|^{-Q}$ away from e, then $\mu - T$ is a finite linear combination of derivatives of the delta function at e. Using the dilations δ_t, it is simple to show that if μ defined a bounded convolution operator on $L_2(G)$ then $\mu - T$ would be a multiple of the delta function at $\{e\}$ (no derivatives could occur). This in turn would imply that T was bounded on L_2. Thus it is enough to show that convolution by T is not bounded on L_2 .

To prove the unboundedness of T , we may exploit the non-integrability of the function $|x|^{-Q}$ either at e or at ∞ . Since we have already gained some control over the singularity at e , it is easier to use the singularity at ∞ . Take a function $f \in C^{\infty}(G)$ such that $f \geq o$ and

$$f(x) = |x|^{-Q/2} (\log |x|)^{-1} \quad \text{when} \quad |x| \geq 2 .$$

By the integration formula of § 1.2 , we have

$$\int_G |f(x)|^2 \, dx \leq C_1 + C_2 \int_2^{\infty} \frac{dt}{t(\log t)^2} < \infty .$$

Thus $f \in L_2(G)$. Furthermore, by the mean value theorem, if $|u| \geq |v|$, then

$$|f(u) - f(v)| \leq C(|u| - |v|) \, |v|^{-(Q+2)/2} .$$

But we know from Corollaries I.3.1 and III.1.1 that

$$\begin{aligned} ||x| - |xy|| &\leq C \, |xy-x| \\ &\leq C \, (|xy-x-y| + |y|) \\ &\leq C \, (|x|^{1-a}|y|^a + |x|^a|y|^{1-a} + |y|) , \end{aligned}$$

where $1/a$ is the length of the filtration on G . It follows from these estimates that

$$|f(xy) - f(x)|^2 \leq C \, |x|^{-Q-2a} \, |y|^{2a} ,$$

whenever $|x| \geq A \, |y|$, and $|y| \leq 1$. Integrating this inequality, we get the following estimate for the L_2 modulus of continuity of right translations on f :

(**) $$\|R_y f - f\|_{L_2} \leq C \, |y|^a ,$$

if $|y| \leq 1$, where $R_y f(x) = f(xy)$.

To show that right convolution by T is unbounded, write $T = \sigma + \rho$, as in (*). Then

$$f * \sigma = \int_{|y| \leq 1} [R_y f - f] \frac{dy}{|y|^Q} ,$$

so it is evident from (**) that $f * \sigma \in L_2(G)$. On the other hand,

$$f * \rho(x) = \int_{|y| \geq 1} f(xy) \frac{dy}{|y|^Q} .$$

But by the estimates above, if $\varepsilon > 0$ is small, then

$$|y| \leq \varepsilon |x| \implies |xy| \geq \varepsilon |x| .$$

Hence if $|x| \geq 1/\varepsilon$, then since $f \geq 0$, we have

$$\begin{aligned} f * \rho(x) &\geq \int_{1 \leq |y| \leq \varepsilon |x|} f(xy) \frac{dy}{|y|^Q} \\ &\geq C \, |x|^{-Q/2} (\log |x|)^{-1} \int_1^{\varepsilon |x|} \frac{dy}{|y|^Q} \\ &\geq C \, |x|^{-Q/2} , \end{aligned}$$

by the integration formula of § 1.2 . This makes it clear that $f * \rho \notin L^2(G)$, and proves the theorem.

Remark In the proof just outlined, we used $f \in C^\infty(G)$ which does not have compact support. To make the proof complete, one must show that there is a sequence of truncations f_n of f, smooth and compactly supported, such that $\{f_n * \sigma\}$ is bounded in L_2 but $\{f_n * \rho\}$ is unbounded in L_2. This can be done using the same estimates (cf. comments and references).

5.2 Filtered nilpotent groups Let $\underline{g}$ be a real, nilpotent Lie algebra, and let F be any positive filtration on $\underline{g}$, as in § I.2.1 . Let $\{\delta_t\}$ and

$|x|$ be dilations and a homogeneous norm, respectively, which are compatible with the filtration F, as in § I.3.1. (Thus δ_t is not an exact Lie algebra automorphism, in general.) Denote by G the vector space $\underline{g}$ with Lie group structure given by the Campbell-Hausdorff formula.

In § I.3.3 we found that near infinity, the group structure of G was asymptotic to the group structure associated with the graded Lie algebra $gr(\underline{g})$. Thus we might expect that the results of § 5.1 should also be valid on G, provided we deal with kernels which are only singular (i.e. not integrable) at infinity. This is the content of the next theorem. (Q = homogeneous dimension of $\underline{g}$ relative to the given filtration.)

Theorem Let k be a locally integrable function on G which is C^1 and homogeneous of degree $-Q + is$ on a neighborhood of infinity. (If $s = o$, assume also that k has mean-value zero on a neighborhood of infinity.) Then the convolution operator $f \mapsto f * k$ is bounded on $L_p(G)$, $1 < p < \infty$.

Proof The operator in question has kernel $K(x,y) = k(y^{-1}x)$. To prove boundedness on L_2, we use the method of decomposition in § 4.2, relative to the distance function $\rho(x,y) = |x^{-1}y|$. Since k is assumed to be locally integrable, it is only necessary in this case to estimate the norm of the operators $T_j T_\ell^*$ and $T_j^* T_\ell$ for j,ℓ large and positive (notation as in § 4.2).

Going back to Example 4 and Remark 2 of § 2.1, Lemma 3.1, and Remarks 1, 3 of § 4.2, we find that the proof of Lemma 4.2 is still valid in this context. The essential point is that the estimates of Lemma 4.2 only involve $K(x,z)$, $K(x,y)$ and $K(y,z)$ for x,y,z such that

$$R^j \leq \rho(x,z) \leq R^{j+1}, \quad R^\ell \leq \rho(y,z) \leq R^{\ell+1}.$$

When j and ℓ are sufficiently large, then these inequalities imply that $\rho(x,y)$ is also large (§ 2.1, Remark 2), and hence ρ and K satisfy the same estimates as in the graded nilpotent case. This proves L_2 boundedness. Boundedness on L_p,

$1 < p < \infty$, follows as in Theorem 4.3 .

5.3 Group germs Let λ be a partial homomorphism from a graded nilpotent Lie algebra V into the Lie algebra of C^∞ vector fields on a manifold M . We assume that for all $x \in M$, the map

$$v \mapsto \lambda(v)_x$$

is a linear isomorphism from V onto the tangent space to M at x . We know from § II.3.1 that M can be covered by open sets X , for which the map

$$\theta : X \times X \to V$$

defined implicitly by the identity

$$e^{\lambda(\theta(x,y))} x = y ,$$

exists and satisfies

1) θ is C^∞ ;

2) For each x , the partial maps

$$y \mapsto \theta(x,y)$$

are diffeomorphisms of X onto an open subset of V .

Let $\rho(x,y) = |\theta(x,y)|$, and let μ be any measure on X which in local coordinates has smooth, strictly positive density relative to Lebesgue measure. Then (X,ρ,μ) is a space of homogeneous type (cf. § 2.1 Example 3, and § 2.2 Example 2). Note that the map θ satisfies the identities

3) $\theta(x,y) = -\theta(y,x)$;

4) $\theta(x,e^{\lambda(u)}x) = u$,

provided u is near 0 in V , so that $e^{\lambda(u)}x \in X$.

In order to analyse the interplay between vector fields on V and vector fields on X, we introduce the notation

$$\varphi(x;u) = \varphi(e^{\lambda(u)}x) ,$$

if φ is a function on M. We take X sufficiently small so that $\varphi(x;u)$ is defined for $x \in X$ and u lying in a fixed neighborhood Ω of 0 in V. If $\varphi \in C_c^\infty(X)$, then $\varphi(x;u)$ is C^∞ on $X \times \Omega$. In particular, if we start with a function f on V, and define $\varphi(x) = f(\theta(y,x))$, for y fixed, then by 4) we have

$$\varphi(y;u) = f(u) .$$

Recall from the lifting theorem of Chapter II that if $w \in V_m$ and $x \in X$, then there is a vector field $T_{x,w}$ on Ω such that

(I) $$(\lambda(w)\varphi)(x;u) = dR(w)\ \varphi(x;u) + T_{x,w}\ \varphi(x;u)$$

(II) $$T_{x,w} \text{ is of order } \leq m-1 \text{ at } u=o .$$

Here dR denotes the right regular representation of the Lie algebra V (cf. § II.2.2), and $\varphi \in C^\infty(M)$. Note that in the present situation the vector field $T_{x,w}$ is uniquely defined by property (I). The point of the lifting theorem in this case is that (II) holds, i.e. in the exponential coordinates centered at x, the vector field $\lambda(w)$ is "well-approximated" by the left-invariant vector field $dR(w)$. In particular, if φ is of the form $\varphi(x) = f(\theta(y,x))$, then we obtain from (4) and (I) the relation

(III) $$(\lambda(w)\varphi)(x) = (dR(w)f)(\theta(y,x)) + T_{y,w}f(\theta(y,x)) .$$

With these preliminaries settled, we now consider operators on X of the form

$$K\varphi(x) = \int_X \varphi(y)\ K(x,y,\ \theta(y,x))\ d\mu(y) ,$$

where $K(x,y,u)$ is a function on $X \times X \times V$, smooth in x and y, and having a prescribed singularity in u at $u = o$. Such an operator can be viewed as a superposition of multiplication operators and "approximate" right-convolution operators (Recall that in the group case, $\theta(y,x) = y^{-1}x)$. The classes of kernels we shall consider are the following:

Definition 1. A function K on $X \times X$ is a kernel of type $s \geq o$ if for every positive integer m we can write

$$K(x,y) = \sum_{i=1}^{N} a_i(x)\, k_i(\theta(y,x))\, b_i(y) + E_m(x,y) ,$$

where

(a) The functions $a_i, b_i \in C_c^\infty(X)$;

(b) The functions $k_i(u)$ are C^∞ on $V \sim \{o\}$, and homogeneous of degree $\alpha_i \geq s - Q$ (with mean-value zero if $s = 0$ and $\alpha_i = -Q$) ;

(c) The remainder term $E_m \in C_c^m(X\times X)$.

(Here Q = homogeneous dimension of V. The number N is allowed to vary with m, of course.)

Theorem 1 If K is a kernel of type $s \geq o$, then the operator

$$A\,\varphi(x) = \int_X K(x,y)\, \varphi(y)\, d\mu(y) ,$$

initially defined on $C_c^\infty(X)$, extends to a continuous operator on $L_p(X)$, for $1 < p < \infty$. (The integral is taken in the principal-value sense when $s = o$.)

Proof We evidently have

$$|K(x,y)| \leq C\, \rho(x,y)^{-Q+s} .$$

In the case $s > o$, this inequality, together with the compact support of K and Lemma 2.2, gives the estimates

$$\begin{cases} \sup\limits_{x} \int\limits_{X} |K(x,y)|\, d\mu(y) < \infty \\ \\ \sup\limits_{y} \int\limits_{X} |K(x,y)|\, d\mu(x) < \infty \quad . \end{cases}$$

Using Hölder's inequality, one then shows easily that A is bounded on L_p, $1 \leq p < \infty$ (cf. proof of Lemma 4.2, for the case $p = 2$). The case $s = o$ follows from Lemma 3.1 and Theorem 4.3 .

Definition 2. A linear operator A mapping $C_c^\infty(X)$ into functions on X is of *type* s, $s > o$, if

$$A\,\varphi(x) = \int\limits_X K(x,y)\, \varphi(y)\, d\mu(y) ,$$

where K is a kernel of type s. If $s = o$, we say A is of *type* 0 if

$$A\,\varphi(x) = PV \int\limits_X K(x,y)\, \varphi(y)\, d\mu(y) + a(x)\, \varphi(x) ,$$

where K is a kernel of type 0 and $a \in C_c^\infty(X)$.

By Theorem 1, an operator A of type $s \geq o$ extends to a continuous operator from $L_p(X)$ to $L_p(X)$, for $1 < p < \infty$. If $f \in L_p(X)$ and D is a differential operator on X with C^∞ coefficients, then DAf will denote the distribution $D(Af)$ on X .

Remarks 1. If A is an operator of type $s \geq o$, then A maps $C^\infty(X)$ into $C_c^\infty(X)$. Indeed, suppose $\varphi \in C_c^\infty(X)$. Then we can write

$$\int\limits_X k(\theta(y,x))\, \varphi(y)\, d\mu(y) = \int\limits_\Omega k(v)\, \varphi(x;-v)\, J(x,v)\, dv ,$$

where $J(x,v)\, dv$ is the image of the measure $d\mu(y)$ under the map $y \mapsto \theta(y,x)$. By assumption, J is a smooth function, so the right side of this formula is a C^∞ function of x . The assertion follows from this calculation and the definition of a kernel of type s .

2. The space of operators of type s is invariant under left and right multiplication by $C^\infty(X)$, and also under transposition.

3. If $f \in C^\infty(X\times\Omega)$ and vanishes to order $m \geq o$ at $u = o$ (as measured by the homogeneous norm on V , cf. § I.1.3) , and if K is a kernel of type s , then $K(x,y)\ f(x,\theta(y,x))$ is a kernel of type $s+m$ (Expand $f(x,u)$ in a Taylor series in u .)

To treat the interaction between operators defined by kernels of type s and differential operators on X we introduce the following filtration:

Definition 3. Let $DO(\lambda)_m$ be the module of differential operators on X which is spanned over $C^\infty(X)$ by 1 and the operators of the form

$$\lambda(w_1)\ \cdots\ \lambda(w_k)\ ,$$

where $w_i \in V$ is homogeneous of degree n_i , and

$$n_1 + \cdots + n_k \leq m\ .$$

We shall refer to elements of $DO(\lambda)_m$ as differential operators of λ-degree $\leq m$. These modules are invariant under transposition (relative to the pairing given by integration over M .)

Theorem 2 If A is an operator of type $s > o$, and $D \in DO(\lambda)_m$, with $m \leq s$, then there exist operators A_1 and A_2 of type $s - m$, such that

$$\begin{cases} DA\ \varphi = A_1\ \varphi \\ AD\ \varphi = A_2\ \varphi \end{cases}$$

for all $\varphi \in C_c^\infty(X)$.

Proof It is obviously sufficient to consider the case $D = \lambda(w)$ and A an operator with kernel $K(x,y) = a(x)\ k(\theta(y,x))\ b(y)$, where $w \in V$ is homogeneous of degree m , and k is a C^∞ function on $V \sim \{o\}$, homogeneous of degree $s - Q$.

Denote by $K_{(1)}$ the function obtained by applying the vector field D to the function $x \mapsto k(\theta(y,x))$. By formula (III) we can write

$$(*) \qquad K_{(1)}(x,y) = dR(w)\, k(\theta(y,x)) + T_{y,w}\, k(\theta(y,x)) \,.$$

Since the vector field $T_{y,w}$ on Ω is of order $\leq m - 1$, it can be expressed as a sum of terms of the form

$$a(y,u)\, D_v \,,$$

where a is a C^∞ function which vanishes to order $\geq n - m + 1$ at $u = o$, and $v \in V$ is homogeneous of degree n . (cf. Chapter I, § 1.3). Hence the function $T_{y,w}\, k(u)$ is a sum of terms

$$a(y,u)\, k_1(u) \,,$$

with a as above and k_1 homogeneous of degree $s - n - Q$. Taking the Taylor expansion in u , we find that the function $T_{y,w}\, k(\theta(y,x))$ is a kernel of type $s - m + 1$.

To analyse the leading term in $(*)$, set

$$F = dR(w)\, k$$

where the derivatives are taken in the distribution sense by identifying k with the distribution $k(u)\, du$. Then as a distribution, F is homogeneous of degree $s - m - Q$. If $s > m$, the pointwise derivative

$$f(u) = \left.\frac{d}{dt}\right|_{t=o} k(u(tw)) \,, \qquad u \neq 0 \,,$$

is a homogeneous function of degree $s - m - Q$, and hence is locally integrable. It follows that the distribution

$$F = f(u)\, du$$

in this case.

In the case $s = m$, the function $f(u)$ is homogeneous of degree $-Q$. We claim that f has mean-value zero, and that

$$(**) \qquad F = PV(f) + C\delta ,$$

where C is a constant, and δ is the delta function at 0. To prove this, we first choose the constant β so that the mean-value of $g(u) = f(u) - \beta|u|^{-Q}$ is zero. The distribution

$$G = F - PV(g)$$

is then homogeneous of degree $-Q$, and away from 0 coincides with the function

$$\beta|u|^{-Q}\, du .$$

But this implies that G is a finite linear combination of the distribution βT and derivatives of the delta function, where T is the distribution

$$< T,\varphi > = \int_{|u|\leq 1} [\varphi(u)-\varphi(o)] \frac{du}{|u|^Q} + \int_{|u|>1} \varphi(u) \frac{du}{|u|^Q} .$$

Under dilations, however, T transforms by

$$< T,\varphi o \delta_t > = < T,\varphi > + C \log(t)\, \varphi(o) ,$$

with $C \neq 0$. Hence G is not homogeneous of degree $-Q$ unless $\beta = 0$. Thus f has mean-value zero, and by homogeneity we obtain $(**)$.

The foregoing analysis thus proves that DA is an operator with kernel

$$a(x)\, f(\theta(y,x))\, b(y) ,$$

modulo operators of type $s - m + 1$ (plus a term $C\, a(x)\, b(y)\, \delta_x(y)$ when $s = m$, where δ_x is the delta function at x).

Taking transposes, we obtain the same conclusion for AD, Q.E.D.

5.4 Boundedness on Sobolev spaces In this section we want to establish the smoothing properties of the class of integral operators of type s introduced in the previous section. For this purpose we introduce the following class of "Sobolev spaces." We continue the assumptions and notation of § 5.3 , and assume that the set X has compact closure in M , and the measure μ is smooth on a neighborhood of X .

Pick a graded basis $\{w_\alpha\}$ for V , and denote by $|\alpha|$ the degree of homogeneity of w_α . If $I = \{\alpha_1,\ldots,\alpha_n\}$ is an ordered collection of indices, set $|I| = |\alpha_1| +\cdots+ |\alpha_n|$ and define

$$D_I = \lambda(w_{\alpha_1}) \cdots \lambda(w_{\alpha_n}) \quad .$$

Definition If $1 < p < \infty$ and m is a non-negative integer, then the space $S_m^p(X)$ consists of all functions $f \in L^p(X)$ such that $D_I f \in L^p(X)$ for all I with $|I| \leq m$. ($D_I f$ is the distribution derivative, using the duality defined by the measure μ .) . Set

$$\|f\|_{p,m} = \sum_{|I|\leq m} \|D_I f\|_{L_p(X)} \quad .$$

As an immediate consequence of Theorems 1 and 2 of § 5.3 , we have

Theorem 1 If A is an operator of type m, m a non-negative integer, then

$$A : L^p(X) \to S_m^p(X)$$

continuously, for $1 < p < \infty$.

We would like to extend this result, and show that A of type m maps S_k^p continuously into S_{k+m}^p for all k . For this we will need the following commutation formula:

Theorem 2 Assume that the graded Lie algebra V is generated by its elements of degree one. Let A be an integral operator of type $s \geq o$ on X and let $D \in DO(\lambda)_m$. Then there exist operators A_i of type s and differential operators $D_i \in DO(\lambda)_m$ such that

$$DA = \sum_{i=1}^{n} A_i D_i \quad .$$

Theorem 2 will follow easily from the following more general result, which does not require that V be generated by its elements of degree one. To state this result, we choose a graded basis $\{w_\alpha\}$ for V as above. Assume that the length of the gradation on V is r .

Theorem 2' Let A be an integral operator of type $s \geq o$ on X and let $w \in V$ be homogeneous of degree m . Then there exist integral operators A_0 of type s , A_α of type $s + |\alpha| - m$, and R_α of type $s + r - m + 1$, such that

$$\lambda(w)\, A = A\, \lambda(w) + A_0 + \sum_{|\alpha|>m} A_\alpha\, \lambda(w_\alpha) + \sum_{\alpha} R_\alpha\, \lambda(w_\alpha) \; .$$

Remark Suppose $m = 1$. Then R_α is of type $s + r$. Since $\lambda(w_\alpha) \in DO(\lambda)_r$ for all α , it follows from Theorem 2 of § 5.3 that the terms involving R_α can be absorbed in A_0 in this case. If we also assume that the elements of degree one generate V , then $\lambda(w_\alpha)$ is the sum of products of $|\alpha|$ vector fields of degree one. By Theorem 2 of § 5.3 , we can compose $|\alpha| - 1$ of these vector fields with A_α , and obtain an operator of type s . This gives Theorem 2 for the case $m = 1$. The general case follows by induction.

Proof of Theorem 2' To clarify the essential idea of the proof, let us consider first the special case in which A is exactly a convolution operator on V , and $\lambda(w)$ is the left-invariant vector field $dR(w)$.

So suppose k is homogeneous of degree $s - Q$, and $\varphi \in C_c^\infty(V)$. Since we are defining the group structure on V using the Campbell-Hausdorf formula, the formula for the adjoint representation of V becomes

$$(\#) \qquad vwv^{-1} = e^{ad(v)}w \ .$$

Thus we can write

$$(\varphi * k)(uw) = \int_V \varphi(uwv^{-1})\, k(v)\, dv$$

$$= \int_V \varphi(uv^{-1}\, e^{ad(v)}w)\, k(v)\, dv \ .$$

Replacing w by tw in this formula and differentiating at $t = o$, we find that

$$(\#\#) \qquad dR(w)(\varphi * k)(u) = \int_V \Phi(uv^{-1}, v)\, k(v)\, dv \ ,$$

where

$$\Phi(u,v) = dR(e^{ad(v)}w)\, \varphi(u) \ .$$

We can expand

$$e^{ad(v)}w = w + \sum_\alpha p_\alpha(v)\, w_\alpha \ ,$$

where p_α are polynomial functions on V (We fix w in this argument). Applying the automorphism δ_t to the identity $(\#)$, we find that

$$p_\alpha(\delta_t v) = t^{|\alpha|-m}\, p_\alpha(v) \ .$$

Since p_α is a polynomial and $p_\alpha(o) = 0$, this implies that $p_\alpha = o$ for $|\alpha| \leq m$. Thus the functions

$$k_\alpha(v) = p_\alpha(v)\, k(v)$$

are homogeneous of degree $s + |\alpha| - m - Q$. Using them, we can write formula $(\#\#)$ as

$$dR(w)(\varphi * k) = (dR(w)\varphi) * k + \sum_{|\alpha|>m} (dR(w_\alpha)\varphi) * k_\alpha .$$

This is the desired commutation formula in this special case.

To treat the general case, we need formulas similar to (#) and (##). Consider first the analogue of the adjoint representation. For v near 0 in V, we have a local one-parameter group of local diffeomorphisms of X given by

$$t \mapsto e^{\lambda(v)} e^{t\lambda(w)} e^{-\lambda(v)} .$$

Denote the generator of this group by $E(v,w)$:

$$E(v,w)\, \varphi(x) = \frac{d}{dt}\Big|_{t=o} \varphi(e^{-\lambda(v)} e^{t\lambda(w)} e^{\lambda(v)} x) ,$$

for $\varphi \in C^\infty(X)$. It is clear from the definition that $E(v,w)$ is a C^∞ vector field on X which depends smoothly on v, when v varies in Ω.

Lemma If $w \in V$ is homogeneous of degree m, then there exist C^∞ functions f_α on $X \times \Omega$ such that

1) $$E(v,w) = \lambda(e^{ad(v)}w) + \sum_\alpha f_\alpha(\cdot,v)\, \lambda(w_\alpha) ;$$

2) the functions $v \mapsto f_\alpha(x,v)$ vanish to order $r - m + 1$ at $v = o$.

Proof of the Lemma: Starting with the formal identity

$$e^X Y e^{-X} = e^{adX} Y ,$$

$(adX(y) = XY-YX)$, one employs the same sort of argument that was used in the proof of Theorem 3.2 of Chapter II. The details are left to the reader.

Completion of proof of theorem: Let $k \in C^\infty(V\sim\{o\})$ be homogeneous of degree $s-Q$ (and with vanishing mean-value, in case $s=o$). Then by Remark 1 in § 5.3, given $\varphi \in C_c^\infty(X)$, we can write

$$(\#\#)' \qquad \int_X k(\theta(y,x))\ \varphi(y)\ d\mu(y) = \int_\Omega \varphi(x;-v)\ k(v)\ J(x,v)\ dv.$$

If we apply the vector field $\lambda(w)$ to the right side, the diffentiations on J will only contribute an operator of type s. By definition, one has

$$\lambda(w)\ \varphi(x;-v) = \frac{d}{dt}\bigg|_{t=o} \varphi(e^{-\lambda(v)}\ e^{t\lambda(w)}x)$$

$$= (E(v,w)\varphi)(e^{-\lambda(v)}x)\ .$$

Using the Lemma above and the formula for $\exp(\mathrm{ad}\ v)\ w$, we find that

$$\lambda(w)\ \varphi(x;-v) = (\lambda(w)\varphi)(x;-v)$$

$$+ \sum_{|\alpha|>m} p_\alpha\ (v)(\lambda(w_\alpha)\varphi)(x;-v)$$

$$+ \sum_\alpha f_\alpha\ (x,v)(\lambda(w_\alpha)\varphi)(x;-v)\ .$$

Substituting this in $(\#\#)'$, we see that the terms involving p_α give operators A_α of type $s + |\alpha| - m$, as in the case of an exact convolution operator. The additional terms involving the functions f_α contribute operators R_α of type $s + r - m + 1$, by Remark 3 in § 5.3 . This proves Theorem 2' in the general case.

<u>Corollary</u> Assume that the Lie algebra V is generated by its elements of degree one. Let A be an integral operator of type $s \geq o$ on X, with s an integer. Then

$$A : S^p_m(X) \to S^p_{m+s}(X)$$

continuously, for $1 < p < \infty$ and $m = 0,1,2,\cdots$.

<u>Proof</u> Let $D \in DO(\lambda)_{m+s}$. By the generating condition, we can write D as a sum of products D_1D_2, with $D \in DO(\lambda)_s$ and $D_2 \in DO(\lambda)_m$. Using Theorem 2 of this section and Theorem 2 of § 5.3 , we find that

$$DA = \Sigma A_i D_i ,$$

with A_i operators of type 0 and $D_i \in DO(\lambda)_m$. Since the operators A_i, D_i and their transposes map $C_c^\infty(X)$ into itself, it follows that this operator equation is valid not only on $C_c^\infty(X)$, but also on $S_m^p(X)$. Thus if $f \in S_m^p(X)$, then $Af \in L_p(X)$ and the distribution derivative $D(Af)$ is the L_p function

$$\Sigma A_i (D_i f) .$$

This completes the proof.

Comments and references for Chapter III

§ 1.1 See Folland [2] and Folland-Stein [1] for further information about homogeneous functions and distributions. The proof of Lemma 1.1 is adapted from Korânyi-Vâgi [1].

§ 1.2 These integral formulas appear in Knapp-Stein [1]. A differential-geometric construction of the fibering of Lebesgue measure by the "spheres" $\{|x| = r\}$ is given in Cotlar-Sandosky [1].

§ 2.1-2.2 The presentation here is a synthesis of the treatments in Koranyi-Vagi [1], Knapp-Stein [1], Folland-Stein [1] and Rothschild-Stein [1]. In particular, Korânyi-Vâgi were the first to emphasize the role played by the "Lipschitz-condition" 4) on the distance function. The map θ in § 2.1, Example 3 was introduced by Folland-Stein [1]. The verification that the associated distance function satisfies axiom 4 was done by Rothschild-Stein [1]; cf. Goodman [7]. If (X, ρ, μ) is a space of homogeneous type, in the sense of § 2.2, then it also satisfies the axioms of Chapter III of Coiffman-Weiss [1], by virtue of Lemma 2.2. The additional conditions that we have imposed which are not used by Coiffman-Weiss are the Lipschitz condition (4) on the distance function, and the logarithmic estimate (*) relating the measure and the distance function. In return, we are able to prove L_p-boundedness of singular integral operators, while they must assume an a-priori L_2 estimate (or prove L_2 boundedness via harmonic analysis, in applications).

§ 2.3 This is adapted from Korânyi-Vâgi [1].

§ 3.1 The same references as in § 2.1-2.2. Our goal in this axiomatic formulation is to isolate the *a priori* information necessary for proving boundedness of singular integrals. For example, Lemma 3.1 can be generalized to include kernels of the form $k(x, \theta(x,y))$, where $k(x,v)$ is C^1 on $X \times (V \sim \{o\})$ and homogeneous of degree $-Q$ in v, with vanishing mean-value. Kernels of this sort naturally occur in the generalizations of the results of Chapter IV, § 3 concerning hypoelliptic operators (cf. Rothschild-Stein [1]).

§ 3.2 This is adapted from Knapp-Stein [1].

§ 4.1 The results of this section go back to Cotlar, in connection with estimates for the classical Hilbert transform. cf. Knapp-Stein [1] and Coiffman-Weiss [1], Chapter VI.

§ 4.2 These estimates are taken from Knapp-Stein [1], Folland-Stein [1] and Rothschild-Stein [1], but adapted to the present axiomatic formulation.

§ 4.3 For the proof of L_p boundedness, $1 < p < 2$ as a consequence of estimate (**) and L_2 boundedness, cf. Coiffman-Weiss [1], Chapter III.

§ 5.1 These results are due to Knapp-Stein [1]. The proof of the "un-boundedness" theorem given here is taken from Goodman [5]. Strichartz [1] has studied singular integrals via the (additive) Fourier transform on certain nil-potent groups.

§ 5.2 These results are new. It would be interesting to extend the comparsion theorem in § 3.3 of Chapter I to a comparison between the operators with kernels $k(y^{-1} x)$ and $k(y^{-1} * x)$, where $*$ means multiplication relative to the graded structure, and k satisfies the conditions of Theorem 5.2.

§ 5.3 The results here are taken from Folland-Stein [1] and Rothschild-Stein [1], reformulated in the context of Chapter II.

§ 5.4 The definition of the chain of Sobolev spaces is adapted from Folland-Stein [1], Folland [2], and Rothschild-Stein [1]. Theorem 2' is stated by

Rothschild-Stein. The proof here, based on the adjoint representation ,is new, as is the Lemma. The Corollary was proved by Folland [2] in the context of a "stratified" nilpotent group (a graded group generated by its elements of degree one). For comparisons between these Sobolev spaces and the usual Sobolev spaces, and for the corresponding Lipschitz spaces, cf. Folland-Stein [1], Folland [2], and Rothschild-Stein [1].

In this chapter we have restricted attention to operators on scalar-valued functions, to minimize the notational burden. Everything works equally well for functions with values in a Hilbert space, and operator-valued kernels. This generalization will be used in Chapter IV, § 1, without further mention (cf. Knapp-Stein [1]).

Chapter IV

Applications

In this chapter we apply the results of the previous chapters to three areas of analysis. The first is the study of irreducibility and equivalences among principal series representations for real-rank one semi-simple Lie groups. In the so-called "non-compact picture", these representations act on $L_2(V)$, V a nilpotent group. The "intertwining integrals" are singular integral operators on V of the type studied in Chapter III.

The second application is the use of non-commutative harmonic analysis on the Heisenberg group to study the Hardy space H^2 on a Siegel domain of type II. The orthogonal projection onto the space of L^2 boundary values of H^2 functions is a singular integral operator, and we calculate its operator-valued Fourier transform. (In this case the boundedness of this operator on L_2 can be proved using the Plancherel theorem for the Heisenberg group.) The Szegö kernel, which reproduces a holomorphic function from its boundary values, is calculated using the Fourier inversion formula on the Heisenberg group.

The goal of the third section is to establish precise regularity properties for certain hypoelliptic differential operators associated with transitive Lie algebras of vector fields. This involves using the full machinery of Chapters II and III. The basic idea, however, is quite simple. Using the lifting theorem, one reduces the problem to the consideration of "approximately invariant" operators on a graded nilpotent group. The corresponding "exactly invariant" operators, which are required to be suitably homogeneous under dilations, have homogeneous fundamental solutions. Approximate fundamental solutions for the original operators are then constructed using the group germ generated by the vector fields and the homogeneous fundamental solutions. The resulting integral operators are of the type studied in Chapter III. The boundedness of these operators on various function spaces yields the desired regularity properties of the original differential operators.

§ 1. Intertwining operators

1.1 Bruhat decomposition and integral formulas Let G be a semi-simple Lie group with finite center. The Iwasawa decomposition of G is

$$G = KAN ,$$

where N is nilpotent, $A \cong \mathbb{R}^{\ell}$ is a vector group normalizing N, and K is a maximal compact subgroup of G. The integer ℓ is the real rank of G. We shall restrict our attention to the case $\ell = 1$. In this case the group N is either commutative or two-step nilpotent, and the action of $Ad(A)$ on N will furnish a group of dilations.

Let M and M' denote the centralizer and the normalizer of A in K, respectively. Then M normalizes N, and

$$B = MAN$$

is a closed subgroup of G. Assuming that real-rank $(G) = 1$, one knows that

$$\# (M'/M) = 2 .$$

Pick $w \in M'$ with $w \notin M$. Then $w^2 \in M$, and

$$w\, a\, w^{-1} = a^{-1} \quad , \quad a \in A$$

$$w\, M\, w^{-1} = M \quad .$$

We define $V = w\, N\, w^{-1}$. Thus V is a nilpotent group isomorphic to N, and

$$w\, B\, w^{-1} = MAV .$$

The map from the product manifold $B \times V$ to G given by $(b,v) \mapsto bv$ is a diffeomorphism onto an open subset of G whose complement has Haar measure zero. The Bruhat decomposition asserts that G is the disjoint union of B double cosets:

$$G = (B\, w\, B) \cup B$$

Multiplying on the right by w, we can write this decomposition as

$$G = (BV) \cup (Bw) .$$

Thus if $g \in G$ and $g \notin Bw$, then there exist unique elements $m \in M$, $a \in A$, $n \in N$, $v \in V$ such that

$$g = man\, v$$

We shall write $m = m(g)$, $a = a(g)$, $v = v(g)$. Then the maps $g \mapsto m(g)$, $g \mapsto a(g)$, $g \mapsto v(g)$ are smooth from the open set BV onto M, A, V respectively.

<u>Example</u> Let $G = SL(2,\mathbb{R})$. We may take

$$K = \left\{ \begin{bmatrix} \cos\theta & \sin\theta \\ -\sin\theta & \cos\theta \end{bmatrix} : \theta \in \mathbb{R} \right\}$$

$$A = \left\{ \begin{bmatrix} a & o \\ o & a^{-1} \end{bmatrix} : a > o \right\}$$

$$N = \left\{ \begin{bmatrix} 1 & x \\ o & 1 \end{bmatrix} : x \in \mathbb{R} \right\} .$$

Then $M = \{\pm I\}$, and we can take

$$w = \begin{bmatrix} o & 1 \\ -1 & o \end{bmatrix}$$

We have

$$B = \left\{ \begin{bmatrix} a & b \\ o & a^{-1} \end{bmatrix} : a \in \mathbb{R} \sim \{o\} \right\}$$

$$V = \left\{ \begin{bmatrix} 1 & o \\ y & 1 \end{bmatrix} : y \in \mathbb{R} \right\}$$

Given

$$g = \begin{bmatrix} a & b \\ c & d \end{bmatrix} \in SL(2,\mathbb{R}) ,$$

we have $g \in BV \Longleftrightarrow d \neq o$, and in this case

$$m(g) = \mathrm{sgn}(d)\ I\ , \quad a(g) = \begin{bmatrix} |d|^{-1} & o \\ o & |d| \end{bmatrix}, \quad v(g) = \begin{bmatrix} 1 & o \\ c/d & 1 \end{bmatrix}.$$

Note that

$$Bw = \left\{ \begin{bmatrix} b & -a \\ a^{-1} & o \end{bmatrix} : a,b \in \mathbb{R}\ ,\ a \neq o \right\} .$$

In terms of the Bruhat decomposition, we have the following integral formulas:

Lemma Let dm, da, dn, dv denote Haar measures on M, A, N, V respectively (all these groups are unimodular). Then

(i) $$\iiint_{MAN} f(man)\ dm\ da\ dn \qquad \text{is a}$$

left Haar integral on $B = MAN$;

(ii) $$\int_B f(b\ man)\ d_\ell b = \mu(a) \int_B f(b)\ d_\ell b\ ,$$

where $d_\ell b$ denotes left Haar measure on B , and

$$\mu(a) = \mathrm{Det}\ (\mathrm{Ad}(a)|_{\underline{n}})\ ;$$

(iii) $$\int_B \int_V f(bv)\ d_\ell b\ dv \qquad \text{is a}$$

Haar integral on G .

Proof (i) follows immediately from the normalization and commutation properties of M, A, N . To prove (ii) , recall that via the exponential map, Lebesgue measure on the Lie algebra $\underline{n}$ serves as Haar measure for N . Since $\det\ (\mathrm{Ad}\ m|_{\underline{n}}) = \det\ (\mathrm{Ad}\ n|_{\underline{n}}) = 1$, we obtain (ii) from (i) and the change of Lebesgue measure under linear transformations.

The proof of (iii) requires a reversal of point of view. We start with a

Haar measure dg on G, and we use dg to define a Haar integral on the direct product group $B \times V$. Namely, we consider the integral

$$I(f) = \int_G f(b(g), v(g))\, dg \ .$$

(f continuous with compact support on $B \times V$.) Here $g = b(g)\, v(g)$ for $g \in BV$, and we note that

$$b_1\, gv_1 = b(b_1 g)\, v(gv_1) \ .$$

Hence $I(f)$ is invariant under left translations by B and right translations by V on $B \times V$. Since V is unimodular, the left Haar measure on $B \times V$ is $d_\ell b\, dv$, so by uniqueness of Haar measure, we must have

$$I(f) = \int_B \int_V f(b,v)\, d_\ell b\, dv \ .$$

Since $G \sim BV$ is of Haar measure zero, this proves (iii).

1.2 Principal series. We continue to assume that G is a semi-simple Lie group of real-rank 1. Let $B = MAN$ as in § 1.1. The finite-dimensional unitary irreducible representations γ of B are all of the form

$$\gamma(man) = \lambda(a)\, \sigma(m) \ ,$$

where λ is a unitary character of A and σ is an irreducible unitary representation of M. (This follows from Engel's theorem: the space of N-fixed vectors for γ is non-trivial and invariant under B, hence is the whole representation space.) Conversely, any such pair (λ,σ) determines an irreducible representation γ of B by this formula. We write $\gamma = (\lambda,\sigma)$, and denote by $H(\sigma)$ the Hilbert space on which σ, and hence γ, acts.

Consider now the unitary representation

$$\pi_\gamma = \operatorname*{Ind}_{B\uparrow G} (\gamma) \ .$$

By definition, π_γ acts on the Hilbert space H_γ of all Borel functions

$$f : G \to H(\sigma)$$

such that for all $man \in B$ and $x \in G$,

(i) $$f(man\ x) = \mu(a)^{1/2}\ \lambda(a)\ \sigma(m)\ f(x)$$

(ii) $$\int_V \|f(y)\|^2\ dy \equiv \|f\|^2 < \infty$$

where dy is Haar measure on V and $\mu(a) = \mathrm{Det}\ (\mathrm{Ad}(a)|_{\underline{n}})$. The action of G is by right translations on H_γ:

$$\pi_\gamma(g)\ f(x) = f(xg) \ .$$

To verify that π_γ is a unitary representation, we need the following result (Recall the decomposition $x = m(x)\ a(x)\ n(x)\ v(x)$):

<u>Lemma</u> For any $g \in G$, the map $y \mapsto v(yg)$, which is defined almost everywhere on V , has inverse $y \mapsto v(yg^{-1})$. Also

$$\int_V f(y)\ dy = \int_V f(v(yg))\ \mu(a(yg))\ dy \ .$$

<u>Example</u> If $G = SL(2,\mathbb{R})$, and

$$y = \begin{bmatrix} 1 & o \\ y & 1 \end{bmatrix}, \quad g = \begin{bmatrix} a & b \\ c & d \end{bmatrix}, \quad \text{then} \quad v(yg) = \begin{bmatrix} 1 & o \\ x & 1 \end{bmatrix},$$

where $x = \frac{ay+c}{by+d}$. The transformation $y \mapsto x$ is defined for all $y \in \mathbb{R}$ except $y = -d/b$. (If $b = o$, i.e. if $g \in MAV$, it is defined for all y .)

We note that $\mu(a) = \alpha^2$, if $a = \begin{bmatrix} \alpha & o \\ o & \alpha^{-1} \end{bmatrix}$.

One calculates $a(yg) = \begin{bmatrix} |by+d|^{-1} & o \\ o & |by+d| \end{bmatrix}$. Hence $\mu(a(yg)) = (by+d)^{-2} = \frac{d}{dy}\left(\frac{ay+c}{by+d}\right)$, which agrees with the formula in the lemma.

Proof of lemma. By definition, the element $v(x)$ is characterised by the property

$$x \in B\, v(x)$$

Hence if $u = v(yg)$, then $y \in B\, ug^{-1}$, so $ug^{-1} \in By$. Since $y \in V$, this gives $v(ug^{-1}) = y$.

To prove the integral formula, we inflate the integral from V to G, by performing an integration over B. Thus if f is continuous with compact support in BV, then

$$\int_B \int_V f(b\, v(yg))\, \mu(a(yg))\, d_\ell b\, dy$$

$$= \int_B \int_V f(b\, n^{-1}\, a(yg)^{-1}\, m^{-1}\, yg)\, \mu(a(yg))\, d_\ell b\, dy ,$$

where m,n are the components of yg in M, N. By part (ii) of lemma 1.1, this integral equals

$$\int_B \int_V f(byg)\, d_\ell b\, dy .$$

By part (iii) of the same lemma, the right translation by g leaves the integral invariant. Specializing to the case $f(bv) = f_1(b)\, f_2(v)$, we obtain the desired formula.

Corollary If $f \in H_\gamma$, $g \in G$, then $\|\pi_\gamma(g)f\| = \|f\|$. Hence $\pi_\gamma(g)$ is unitary.

Proof For $y \in V$, write $yg = ma(yg)\, n\, v(yg)$. Then by transformation property (i) of f and the unitarity of λ, σ, we have

$$\|f(yg)\|^2 = \mu(a(yg))\, \|f(v(yg))\|^2 .$$

Integrating over V and using the lemma, we see that $\pi_\gamma(g)$ is unitary, Q.E.D.

The family $\{\pi_{\lambda,\sigma} : \lambda \in \hat{A},\ \sigma \in \hat{M}\}$ of unitary representations of G is called the unitary principal series of representations.

1.3 *Intertwining operators* By means of the principal series, we have defined a mapping $\gamma = (\lambda,\sigma) \mapsto \pi_\gamma$ from $\hat{A} \times \hat{M}$ to the set of unitary representations of G, where $\hat{A}$ and $\hat{M}$ denote the spaces of irreducible unitary representations of A and M. This map can be viewed as a "ramified covering map" from $\hat{M} \times \hat{A}$ onto a subset of $\hat{G}$. The "monodromy group" of this covering is the Weyl group M'/M, and the action of the Weyl group is given by certain singular integral operators on the nilpotent group V.

In more detail, we note first that if $w \in M'$ is a fixed representative for the non-trivial element of M'/M, then we can define an action of M'/M on $\hat{M} \times \hat{A}$ by setting

$$\begin{cases} (w\cdot\lambda)(a) = \lambda(w^{-1}aw) \\ w\cdot\sigma(m) = \sigma(w^{-1}mw) . \end{cases}$$

If $\gamma = (\lambda,\sigma)$, write $w\cdot\gamma = (w\cdot\gamma, w\cdot\sigma)$. Then by general results of Bruhat, one knows that

(i) $\pi_\gamma \simeq \pi_{\gamma'} \iff \gamma' \simeq \gamma$ or $\gamma' \simeq w\cdot\gamma$

(ii) If $w\cdot\gamma \neq \gamma$, then π_γ is irreducible.

Combining (i) and (ii), we see that on the set $\{\gamma : w\,\gamma \neq \gamma\}$ the map $\gamma \mapsto \pi_\gamma$ carries $\hat{M} \times \hat{A}$ into $\hat{G}$, and is a two-sheeted covering. The "ramification points" are the points $\{\gamma : w\,\gamma \simeq \gamma\}$. Since $(w\cdot\lambda)(a) = \lambda(a^{-1})$, it follows that $\lambda = 1$ if $\gamma = (\lambda,\sigma)$ is a ramification point. We shall use the theory of singular integral operators developed in Chapter III to

(a) construct the unitary operator giving equivalence between π_γ and $\pi_{w\cdot\gamma}$ when $\gamma \neq w\cdot\gamma$;

(b) determine the reducibility or irreducibility of π_γ when $\gamma \simeq w\cdot\gamma$.

From an intuitive point of view, perhaps the most natural starting point is the following formal change-of-variables argument:

Assume $f \in H_\gamma$, and define

$$A(\gamma)\, f(x) = \int_V f(yw^{-1}x)\, dy\ .$$

If $b = man \in B$, then for $y \in V$ we have

$$f(yw^{-1}bx) = f(ym^w\, a^w\, n^w\, w^{-1}x)$$

$$= \mu(a^w)^{1/2}\, \lambda(a^w)\, \sigma(m^w)\, f(y^z\, n^w\, w^{-1}x)\ ,$$

where we write $x^g = g^{-1}ng$, and $z = m^w\, a^w$. But $a^w = a^{-1}$ and $n^w \in V$. Hence assuming that the integral converges, we have

$$A(\gamma)\, f(bx) = \mu(a)^{-1/2}(w\cdot\gamma)(b) \int_V f(y^z\, n^w\, w^{-1}x)\, dy$$

$$= \mu(a)^{1/2}(w\cdot\gamma)(b)\, A(\gamma)\, f(x)\ .$$

Here we have used the fact that for $z = m^w\, a^w$,

$$\int_V f(y^z)\, dy = \int_V f(y^{a^{-1}})\, dy$$

$$= \mu(a) \int_V f(y)\, dy\ .$$

It follows from the above calculation that in a formal sense,

$$A(\gamma) : H_\gamma \to H_{w\cdot\gamma}\ .$$

Obviously $A(\gamma)$ commutes with right translations by elements of G .

To investigate the convergence of the integral $A(\gamma)f$, we use lemma 1.2 to rewrite $A(\gamma)f$ as a convolution integral on V . Namely, we write

$$yw^{-1} = m(yw^{-1})\, a(yw^{-1})\, n\, v(yw^{-1})\ .$$

Then for $f \in H_\gamma$, we have

$$f(yw^{-1}x) = \mu(yw^{-1})^{1/2} \gamma(yw^{-1}) f(v(yw^{-1})x) ,$$

where we have written $\mu(yw^{-1}) = \mu(a(yw^{-1}))$, $\gamma(yw^{-1}) = \lambda(a(yw^{-1})) \sigma(m(yw^{-1}))$. Introduce the notation $\overline{yg} = v(yg)$. Then by lemma 1.2,

$$y = \overline{(\overline{yg})\, g^{-1}} \qquad (y \in V) ,$$

and

$$\int_V \varphi(\overline{yg})\, dy = \int_V \varphi(y)\, \mu((\overline{yg^{-1}})g)^{-1}\, dy ,$$

for any integrable function φ on V and $g \in G$. Using this change of variables in the integral defining $A(\gamma)$, we have the formal convolution integral

$$A(\gamma) f(x) = \int_V \mu(\overline{yw}\, w^{-1})^{-1/2} \gamma(\overline{yw}\, w^{-1}) f(yx)\, dy .$$

We can simplify this last expression by noting that if $y \in V$, then

$$a(\overline{yw}\, w^{-1}) = a(yw)^{-1}$$

$$m(\overline{yw}\, w^{-1}) = m(yw)^{-1}$$

$$a(yw^{-1}) = a(yw) .$$

These identities follow from the inclusions

$$\overline{yw}\, w^{-1} \in m(yw)^{-1} a(yw)^{-1} \mathbf{N}\, y$$

and

$$yw^{-1} \in yw\, M .$$

Thus if K_γ is defined on V by

$$K_\gamma(y) = \mu(yw)^{1/2} \gamma(yw)^{-1}$$

then the operator $A(\gamma)$ is formally expressible as

$$A(\gamma) f(x) = \int_V K_\gamma(y) f(yx)\, dy .$$

<u>Lemma</u> The function K_γ is C^∞ on $V \sim \{e\}$. Furthermore,

$$\begin{cases} K_\gamma(y^a) = \mu(a)^{-1} \lambda(a)^2 K_\gamma(y) \\ K_\gamma(y^{-1}) = K_{w\gamma}(y)^* \sigma(w^2)^* \end{cases}$$

(Here $y^a = a^{-1}ya$, and $\gamma = (\lambda,\sigma)$.)

<u>Proof</u> If $y \in V$ and $y \neq e$, then $y w \notin Bw$. Hence by the Bruhat decomposition, $y w \in BV$. Thus the maps $y \mapsto m(yw)$ and $y \mapsto a(yw)$ are C^∞ away from e. This gives the smoothness of K_γ.

To establish the transformation properties of K_γ, we recall that $aw = wa^{-1}$, for $a \in A$. Hence

$$y^a w = a^{-1} yw\, a^{-1} \in m(yw) a^{-2}\, a(yw)\, NV$$

so that

$$a(y^a w) = a^{-2}\, a(yw) \quad .$$

This gives the transformation law under $y \mapsto y^a$. Since

$$y\, w \in m(yw)\, a(yw)\, NV \quad ,$$

we have

$$y^{-1} w \in w\, VN\, m(yw)^{-1}\, a(yw)^{-1}\, w \quad .$$

But $w\,V = N\,w$, so that

$$y^{-1}\, w \in \left[w\, m(yw)^{-1}\, w\right] a(yw)\, NV \quad .$$

This shows that

$$\begin{cases} a(y^{-1}\, w) = a(yw) \\ m(y^{-1}\, w) = w\, m(yw)^{-1}\, w \end{cases} ,$$

which gives the transformation law under $y \mapsto y^{-1}$. (Note that $\gamma(y)^{-1} = \gamma(y)^*$, since γ is unitary.)

<u>Example</u> When $G = SL(2,\mathbb{R})$, the characters of A are given by

$$\begin{bmatrix} a & o \\ o & a^{-1} \end{bmatrix} \mapsto a^{\lambda} \quad , \qquad \lambda \in i\mathbb{R} \ ,$$

and $\hat{M} = \{1,\varepsilon\}$, where $\varepsilon(-I) = -1$. If

$$y = \begin{bmatrix} 1 & o \\ y & 1 \end{bmatrix}, \ w = \begin{bmatrix} o & 1 \\ -1 & o \end{bmatrix}, \ \text{then} \ yw = \begin{bmatrix} o & 1 \\ -1 & y \end{bmatrix},$$

so that

$$a(yw) = \begin{bmatrix} |y|^{-1} & o \\ o & |y| \end{bmatrix}, \qquad m(yw) = \mathrm{sgn}(y)\ I \ .$$

If $\gamma_+ = (\lambda,1)$, $\gamma_- = (\lambda,\varepsilon)$ are the corresponding representations of B, then this calculation shows that

$$K_{\gamma_+}(y) = |y|^{-1+\lambda}$$

$$K_{\gamma_-}(y) = \mathrm{sgn}(y)\ |y|^{-1+\lambda} \ .$$

Note that

$$\mu(yw)^{-1/2} = |y| \ .$$

1.4 <u>Boundedness of intertwining operators</u> We shall now apply the results of Chapter III to the kernels K_γ. Identify V with its Lie algebra by the exponential map. Since real-rank $(G) = 1$, it follows from the properties of root systems that

$$V = V_1 \oplus V_2 \ ,$$

$$A = \{a(t) : t \in \mathbb{R}^+\}$$

such that

$$y^{a(t)} = \begin{cases} ty & , \quad y \in V_1 \\ t^2 y & , \quad y \in V_2 \end{cases} .$$

We define dilations $\{\delta_t, t > o\}$ on V by

$$\delta_t \, y = y^{a(t)} .$$

The homogeneous dimension Q of V is then

$$Q = \dim (V_1) + 2 \dim (V_2)$$

Let $\qquad \mu(a) = \operatorname{Det} (\operatorname{Ad}(a)|_{\underline{n}})$, and set

$$|y| = \mu(yw)^{-1/(2Q)} .$$

<u>Lemma</u> $|y|$ is a smooth, symmetric δ_t-homogeneous norm on V , and $|y^m| = |y|$, $|\overline{yw}| = |y|^{-1}$ for $y \in V$, $m \in M$.

<u>Proof</u> As in the proof of Lemma 1.3, we calculate that when $t > o$,

$$a((\delta_t y)w) = a(t)^2 \, a(yw) .$$

Hence $\qquad |\delta_t y| = \mu(a(t))^{-1/Q} \, |y|$.

But $\qquad N = \operatorname{Ad}(w) \, V$, so for $a \in A$ one has

$$\operatorname{Det} (\operatorname{Ad}(a)|_N) = [\operatorname{Det} \operatorname{Ad}(a)|_V]^{-1}$$

Since $\qquad \operatorname{Det} \operatorname{Ad}(a(t))|_V = t^Q$, this shows that

$$|\delta_t y| = t \, |y| , \qquad t > o .$$

The function $y \mapsto a(yw)$ is smooth on $V \sim \{e\}$, and $a(yw) = a(y^{-1}w)$. Thus $|y|$ is a smooth, symmetric norm on V .

The M-invariance of $|y|$ follows from the identity $a(y^m w) = a(yw)$. For the transformation $y \mapsto \overline{yw}$, we have

$$(\overline{yw})w \quad \epsilon \quad M\, a(yw)^{-1}\, NV \ ,$$

and hence $a(\overline{yw})w = a(yw)^{-1}$. Thus $|\overline{yw}| = |y|^{-1}$, Q.E.D.

Suppose $\gamma(ma(t)) = t^{\lambda/2}\,\sigma(m)$, where $\lambda \in i\,\mathbb{R}$. If $\Omega_\gamma(y) = \gamma(wy)^{-1}$, then by lemma 1.3 one sees that Ω_γ is homogeneous of degree λ with respect to the dilations δ_t . Furthermore, the kernel K_γ for the intertwining operator $A(\gamma)$ is given by

$$K_\gamma(y) = \Omega_\gamma(y)\, |y|^{-Q} \qquad ,$$

and is homogeneous of degree $-Q+\lambda$.

Recall that the calculations involving $A(\gamma)$ in § 1.3 ignoved any convergence problems. It is evident from this formula for K_γ that $A(\gamma)$ is formally a singular integral operator of the type treated in Chapter III. The homogeneity and smoothness conditions are satisfied by K_γ for any γ . The mean-value condition, however, is not always satisfied.

Theorem (a) Suppose that $w\gamma \neq \gamma$. Then there exists $R > 1$ such that the integral of K_γ over $\{A \leq |y| \leq AR\}$ is zero for all $A > 0$. The operator

$$A(\gamma)\, f(x) = \text{P.V.} \int_V K_\gamma(y)\, f(yx)\, dy$$

is a non-zero bounded operator from H_γ to $H_{w\gamma}$ which intertwines the representations π_γ and $\pi_{w\gamma}$. Some scalar multiple of $A(\gamma)$ is a unitary operator.

(b) Suppose that $w\gamma \simeq \gamma$. Then the representation π_γ is reducible if and only if the mean-value of the function $y \mapsto \text{tr}\,(\sigma(w)\sigma(yw))$ is zero. In this case π_γ splits as the direct sum of two inequivalent irreducible representations. The projection operator giving the decomposition is a linear combination of I and the operator $\sigma(w)\, A(\gamma)$.

Remarks 1. In part (b) , we are using the fact that when $w\gamma \simeq \gamma$, then we may define an operator $\sigma(w)$ which extends the representation σ from

M to M'. Indeed, by assumption there exists a unitary operator T_o on $H(\sigma)$ such that

$$\sigma(w^{-1}mw) = T_o^{-1}\,\sigma(w)\,T_o \quad .$$

Since $w^2 \in M$, one finds that $T_o^2\,\sigma(w^2)^{-1}$ commutes with $\sigma(m)$, and hence is a scalar. Thus we can choose $\theta \in \mathbb{R}$ such that

$$T = e^{i\theta}\,T_o$$

satisfies

$$T^2 = \sigma(w^2) \quad .$$

We set $\sigma(w) = T$.

2. When $w\gamma \approx \gamma$, then $\lambda = o$, and the operator sending $f \to \sigma(w)^{-1}f$ is a unitary map from H_γ to $H_{w\cdot\gamma}$ which intertwines π_γ and $\pi_{w\cdot\gamma}$.

Proof of theorem We begin by determining the transformation properties of the kernel Ω_γ under the automorphisms $y \mapsto y^m$, $m \in M$. Note that

$$y^m\,w = m^{-1}\,yw\,m^w \quad .$$

Hence the M-component of $y^m\,w$ is $m^{-1}\,m_1\,m^w$, where m_1 is the M-component of yw. We already calculated in Lemma 1.4 that

$$a(y^m\,w) = a(yw) \quad .$$

Hence we obtain the formula

$$\Omega_\gamma\,(y^m) = (w\cdot\sigma)(m^{-1})\,\Omega_\gamma\,(y)\,\sigma(m) \quad .$$

Since $|y| = |y^m|$ and $\mathrm{Det}\,(\mathrm{Ad}(m)|_V) = 1$, we may integrate this formula over a shell $\{a \leq |y| \leq b\}$ and obtain the relation

$$(*) \qquad T_\gamma = (w\cdot\sigma)(m^{-1})\,T_\gamma\,\sigma(m) \quad ,$$

where T_γ = mean-value of the function $|y|^{-\lambda}\,\Omega_\gamma(y)$.

Suppose now that $w\gamma \neq \gamma$. If $\lambda \neq 0$ and $\lambda \in i\mathbb{R}$, then by Lemma III.3.1 there is an $R > 1$ such that the integral of K_γ over $\{A \leq |y| \leq AR\}$ is zero for any $A > 0$. By Lemma 1.3, the adjoint kernel $K_\gamma(y^{-1})^*$ also satisfies this condition, for the same value of R. The smoothness and homogeneity conditions are satisfied by K and K^*. Hence $A(\gamma)$ exists as a bounded operator on $L^2(V)$. If $\lambda = 0$, then $w\sigma \neq \sigma$. Since the mean-value T_γ intertwines $w\cdot\sigma$ and σ, by (*), we must have $T_\gamma = 0$ in this case. Thus Lemma III.3.1 also applies, and $A(\gamma)$ exists as a bounded operator on $L^2(V) \otimes H(\sigma)$.

The representation space of π_γ and $\pi_{w\cdot\gamma}$ can be identified with $L^2(V) \otimes H(\sigma)$, via the map $f \mapsto f|_V$. (This is the so-called "non-compact picture" for the representation.) In this realisation, the subgroup V acts by right translations, and the subgroup MA acts by

$$\pi_\gamma(ma)\, f(y) = \mu(a)^{1/2}\, \gamma(ma)\, f(y^{ma}) \quad .$$

The element w acts by

$$\pi_\gamma(w)\, f(y) = \mu(yw)^{1/2}\, \gamma(yw)\, f(\overline{yw})$$

Since $G = (MAV) \cup (MAV\, w\, V)$, these formulas determine π_γ.

It is obvious that $A(\gamma)$ commutes with right translations by V. By Lemma 1.3 and the calculation above we find that

$$(**) \qquad K_\gamma(y)\, \gamma(ma) = \mu(a)(w\cdot\gamma)(ma)\, K_\gamma(y^{ma}) \quad .$$

Suppose $f \in C_c^\infty(V) \otimes H(\sigma)$. Then $A(\gamma)f$ is given by the absolutely convergent integral

$$A(\gamma)\, f(x) = \int_V K_\gamma(y)\, [f(yx) - f(x)]\, dy \quad ,$$

since K_γ has integral zero over the family of shells $\{R^n \leq |y| \leq R^{n+1}\}$. Using equation (**) and the integration formula

$$\int_V f(y^{ma})\, \mu(a)\, dy = \int_V f(y)\, dy \quad ,$$

we verify easily that if $g \in MAV$, then

$$A(\gamma)\ \pi_\gamma(g)\ f(x) = \pi_{w\cdot\gamma}\ (g)\ A(\gamma)\ f(x) \quad .$$

The proof that $A(\gamma)$ intertwines $\pi_\gamma(w)$ is more delicate. The most conceptual verification seems to be to return to the formula for $A(\gamma)$ as originally given, but take $\text{Re}\ \lambda > o$. Then one proves that the integral defining $A(\gamma)$ now converges absolutely for f in H_γ^∞, where

$$H_\gamma^\infty = \{f \in C^\infty(G,H(\sigma));\ f(\text{man}\ g) = \mu(a)^{1/2}\ \gamma(ma)f(g)\} \ .$$

The same change of variable argument shows that

$$A(\gamma)\ :\ H_\gamma^\infty \to H_{w\cdot\gamma}^\infty \quad ,$$

and $A(\gamma)$ commutes with right translations by G. One proves that as $\text{Re}\ \lambda \searrow o$, $A(\gamma)$ converges to the singular integral operator constructed above. For details we refer to the literature cited at the end of the chapter.

To finish the proof of part (a), we recall that by the results of Bruhat cited earlier, the representation π_γ is irreducible if $w\cdot\gamma \neq \gamma$. Hence $A(\gamma)^* A(\gamma)$ must be a non-zero multiple of the identity operator, since it commutes with π_γ. Similary, $A(\gamma)\ A(\gamma)^*$ is a multiple of the identity operator. Thus with a suitable normalization, $A(\gamma)$ becomes unitary.

In part (b), $w\cdot\gamma \simeq \gamma$ implies that $\lambda = o$ and $w\cdot\sigma \simeq \sigma$. We extend the representation σ to M' as noted in remark 1. Then the calculation at the beginning of the proof shows that

$$(\sigma(w)\ T_\sigma) = \sigma(m)^{-1}(\sigma(w)T_\sigma)\ \sigma(m) \quad ,$$

where T_σ is the mean-value of the matrix function $y \mapsto \sigma(yw)$. Hence $\sigma(w)T_\sigma$ is a scalar multiple of I, since σ is irreducible. We conclude that $T_\gamma = o$ if and only if the mean-value of the function

$$y \mapsto \text{tr}\ (\sigma(w)\ \sigma(yw))$$

is zero.

Suppose this mean-value is zero. The argument above shows that $A(\gamma)$ is a bounded operator from H_γ to $H_{w\gamma}$ which intertwines π_γ and $\pi_{w\gamma}$. The operator $\sigma(w)\,A(\gamma)$ is then a bounded operator from H_γ to H_γ which commutes with π_γ (cf. remark 2 above). On the other hand, the results of Bruhat imply that the order of the Weyl group (two, in this case) always majorizes the number of irreducible components of π_γ. Since $\sigma(w)\,A(\gamma)$ is not the identity operator on $L^2(V) \otimes H(\sigma)$, we conclude that every intertwining operator is a linear combination of $\sigma(w)\,A(\gamma)$ and I, and the intertwining ring for π_γ is two-dimensional (and hence commutative). Thus π_γ splits as $\pi_\gamma^+ \oplus \pi_\gamma^-$, where $\pi_\gamma^\pm$ are irreducible and inequivalent.

It only remains to verify that if the mean value of Ω_γ is <u>not</u> zero, then the intertwining ring for π_γ is trivial. By the results of Bruhat, any intertwining operator T is expressible as left convolution by a distribution on V, and away from $\{e\}$ this distribution is the function $y \mapsto c\sigma(w)\sigma(yw)$, where c is a constant. By the "unboundedness" Theorem III. 5.1, T cannot be a bounded operator unless $c = o$. This implies that T is a multiple of I, Q.E.D.

1.5 <u>Examples</u> Let us illustrate the reducibility criterion of Theorem 1.4. Suppose first that $G = SL(2,\mathbb{R})$. Then $w\cdot\gamma = \gamma$ means that γ is either trivial or else $\gamma = \varepsilon$, where $\varepsilon(\pm I) = \pm 1$. The mean-value zero condition is satisfied only in the second case, and we have in this case

$$K_\varepsilon(y) = 1/c\ , \quad \text{if} \quad y = \begin{pmatrix} 1 & o \\ c & 1 \end{pmatrix} .$$

Thus $A = \pi^{-1}A(\varepsilon)$ is the classical Hilbert transform:

$$A\,f(x) = \text{P.V.}\ \frac{1}{\pi} \int_{-\infty}^{\infty} \frac{f(t)\,dt}{t-x} .$$

After Fourier transformation A becomes multiplication by the function $i\,\mathrm{sgn}(\xi)$, so $A^2 = -I$. The spectral decomposition of A is given by

$$L^2(\mathbb{R}) = H_+^2(\mathbb{R}) \oplus H_-^2(\mathbb{R}) ,$$

where $H_\pm^2(\mathbb{R})$ are the Paley-Wiener spaces of L^2 functions holomorphic in the upper (lower) half plane, with

$$\sup_{y>0} \int_{-\infty}^{\infty} |f(x \pm iy)|^2 \, dy < \infty .$$

The representation π_ε in the non-compact picture is given by

$$\pi_\varepsilon(g)\, f(x) = (bx+d)^{-1}\, f\left(\frac{ax+c}{bx+d}\right) ,$$

if $g = \begin{bmatrix} a & b \\ c & d \end{bmatrix}$. It is evident from the above description that $H_\pm^2$ are invariant under $\pi_\varepsilon(g)$. Theorem 1.4 asserts that the restriction of π_ε to $H_\pm^2$ is irreducible.

As an other example, consider the group $G \subset SL(3,\mathbb{C})$ which leaves invariant the Hermitian form

$$z_2^* z_2 + 2\,\mathrm{Re}(z_1 z_3^*) ,$$

where $(z_1, z_2, z_3) \in \mathbb{C}^3$. (This group is conjugate to the group $SU(2,1)$ which leaves invariant the form $z_1^* z_1 + z_2^* z_2 - z_3^* z_3$.) The subgroups M, A, N, V in this case are the following (all blank matrix entries are zeros):

$$M : \qquad m_\theta = \begin{bmatrix} e^{i\theta} & & \\ & e^{-2i\theta} & \\ & & e^{i\theta} \end{bmatrix} , \qquad \theta \in \mathbb{R}$$

$$A : \qquad a_r = \begin{bmatrix} r & & \\ & 1 & \\ & & r^{-1} \end{bmatrix} , \qquad r > 0$$

$$N : \qquad \exp \begin{bmatrix} 0 & z^* & it \\ & 0 & z \\ & & 0 \end{bmatrix} , \qquad z \in \mathbb{C},\ t \in \mathbb{R}$$

$$V: \qquad \exp\begin{bmatrix} o & & \\ z & o & \\ it & z^* & o \end{bmatrix}, \qquad z \in \mathbb{C},\ t \in \mathbb{R}$$

For the Weyl group representative, we take

$$w = \begin{bmatrix} & & i \\ & 1 & \\ i & & \end{bmatrix}$$

It is then a straightforward calculation to determine the matrices $m(vw)$, $a(vw)$ and $\overline{vw}$, when $v = v(z,t)$, where

$$v(z,t) = \exp\begin{bmatrix} o & & \\ z & o & \\ it/2 & z^* & o \end{bmatrix} = \begin{bmatrix} 1 & & \\ z & 1 & \\ \frac{1}{2}(z^*z+it) & z^* & 1 \end{bmatrix}.$$

($u = \overline{vw}$ is uniquely determined by the property that $(vw)u^{-1}$ is an upper triangular matrix, whose diagonal entries then give $m(vw)$ and $a(vw)$.) One finds that

$$\begin{cases} a(vw) = a_r, & \text{where} \quad r = 2|z^*z+it|^{-1} \\ m(vw) = m_\theta, & \text{where} \quad \theta = \arg[i(z^*z+it)] \\ \overline{vw} = v(\zeta,\tau), & \text{where} \end{cases}$$

$$\zeta = \frac{2iz}{z^*z-it}, \quad \tau = -\frac{4t}{|z^*z-it|^2}.$$

The adjoint action of A on V is given by

$$Ad(a_r)\, v(z,t) = v(rz, r^2t),$$

when $r > o$, and the homogeneous dimension of V is $Q = 4$. The modular function $\mu(a_r) = r^4$, and hence the homogeneous norm on V is given by

$$|v| = \mu(vw)^{-1/8} = \frac{1}{\sqrt{2}}\,|z^*z+it|^{1/2}.$$

Here $\qquad v = v(z,t)$.

The group $M \simeq U(1)$ in this case, and $\hat{M}$ consists of all representations

$$\sigma_n(m_\theta) = e^{-in\theta} , \qquad n \in \mathbb{Z} .$$

The action of w on M is trivial. Let

$$K_n(v) = \mu(vw)^{1/2} \sigma_n(vw)^{-1} , \qquad v \in V .$$

By the formulas above we can write

$$K_n(v) = c_n \frac{\Omega_n(v)}{|v|^4} ,$$

where

$$\Omega_n(v) = \frac{(z^*z+it)^n}{|z^*z+it|^n} , \qquad v = v(z,t) ,$$

and c_n is a non-zero constant. Denote by π_n the representation of G induced by the character $ma \longrightarrow \sigma_n(m)$.

<u>Theorem</u> π_n is reducible $\Longleftrightarrow$ n is even and non-zero.

<u>Proof</u>. We shall show that the mean-value of Ω_n is zero $\Longleftrightarrow$ n is even, $n \neq o$. For this purpose, we write $z^*z+it = re^{i\theta}$. Then using cylindrical coordinates, we can express

$$\int_{a<|v|<b} \frac{\Omega_n(v)}{|v|^4} dv = c \int_a^b \left(\int_0^\pi e^{in\theta} d\theta\right) \frac{dr}{r} .$$

Thus the mean-value of Ω_n is given by the integral

$$\int_0^\pi e^{in\theta} d\theta ,$$

which vanishes precisely for n even , $n \neq o$, Q.E.D.

§ 2 Boundary values of H^2 functions

2.1 Harmonic analysis on the Heisenberg group Let D be the "Siegel domain of type II" introduced in Chapter II, § 4.4. The Heisenberg group G acts simply transitively on the boundary M of D. Using the (non-commutative). Fourier analysis on G, we shall study the boundary values of functions in the Hardy class $H^2(D)$.

We first recall the basic facts concerning harmonic analysis on G. We parametrize G as $\mathbb{R}^n \times \mathbb{R}^n \times \mathbb{R}$, as in § II.4.4, with multiplication

$$(\xi,\eta,\zeta)(\xi',\eta',\zeta') = (\xi+\xi',\eta+\eta',\zeta+\zeta'+\frac{1}{2}(\xi\cdot\eta'-\xi'\cdot\eta)) .$$

For every $\lambda \in \mathbb{R} \sim \{0\}$, there is an irreducible unitary representation π^λ of G acting on $L^2(\mathbb{R}^n)$ by

$$\pi^\lambda(g)\, f(x) = e^{i\lambda(\zeta+\eta\cdot x+\frac{1}{2}\eta\cdot\xi)}\, f(x+\xi) ,$$

where $g = (\xi,\eta,\zeta)$, and $x\cdot y = \Sigma\, x_i y_i$ $(x,y \in \mathbb{R}^n)$.

Given $\varphi \in L^1(G)$, we define the operator $\hat{\varphi}(\lambda)$ on $L^2(\mathbb{R}^n)$ by the operator-valued integral

$$\hat{\varphi}(\lambda) = \int_G \pi^\lambda(g)^* \varphi(g)\, dg .$$

Here dg is Haar measure on G (=Lebesgue measure on $\mathbb{R}^{2n+1}$ in the above coordinates). The Plancherel formula is

$$\int_G |\varphi(g)|^2\, dg = \int_{\mathbb{R}} \|\hat{\varphi}(\lambda)\|_{HS}^2\, d\mu(\lambda) ,$$

where $\|T\|_{HS}^2 = \mathrm{tr}\,(T^*T)$ is the square of the Hilbert-Schmidt norm, and the Plancherel measure

$$d\mu(\lambda) = c_n\, |\lambda|^n\, d\lambda ,$$

with $c_n = (2\pi)^{-n-1}$ and $d\lambda$ Lebesgue measure on $\mathbb{R}$. If we define $L^2(\hat{G})$ to be the Hilbert space of all measurable, operator-valued functions $\lambda \mapsto T(\lambda)$ on $\mathbb{R}$ such that

$$\int_{\mathbb{R}} \|T(\lambda)\|_{HS}^2 \; d\mu(\lambda) < \infty$$

(where $T(\lambda)$ is an operator on $L^2(\mathbb{R}^n))$, then the map $\varphi \mapsto \hat{\varphi}$ extends to a unitary map from $L^2(G)$ onto $L^2(\hat{G})$.

Let ρ be the right regular representation of G on $L^2(G)$:

$$\rho(g)\;\varphi(x) = \varphi(xg)$$

The Fourier transform of $\rho(g)$ is then left multiplication by the operator-valued function $\pi^\lambda(g)$:

$$(\rho(g)\varphi)^\wedge(\lambda) = \pi^\lambda(g)\;\hat{\varphi}(\lambda) .$$

To study the "tangential Cauchy-Riemann" equations satisfied by the boundary values of holomorphic functions, we shall need to extend ρ to a representation of the Lie algebra $\underline{g}$ of G. If π is any unitary representation of (any Lie group) G on a Hilbert space $H(\pi)$, we define the subspace $H^\infty(\pi)$ of C^∞ vectors for π by

$$H^\infty(\pi) = \{v \in H(\pi) : g \mapsto \pi(g)v \text{ is a } C^\infty \text{ function}\} .$$

This is a dense subspace of $H(\pi)$. Given $v \in H^\infty(\pi)$ and $X \in \underline{g}$, we define

$$\pi_\infty(X)\;v = \frac{d}{dt}\Big|_{t=0} \pi(\exp tX)\;v .$$

Then $\pi_\infty(X) : H^\infty(\pi) \to H^\infty(\pi)$, and the map $X \mapsto \pi_\infty(X)$ is a Lie algebra homomorphism from $\underline{g}$ to operators on $H^\infty(\pi)$. Hence it extends uniquely to an associative algebra homomorphism from the complexified universal enveloping algebra $U(\underline{g})$ into the algebra of linear transformations of $H^\infty(\pi)$, which we also denote by π_∞.

Give $H^\infty(\pi)$ the topology defined by the family of semi-norms

$$v \mapsto \|\pi_\infty(T)v\| ,$$

as T ranges over $U(\underline{g})$. Then $H^\infty(\pi)$ is a Fréchet space. We denote by $H^{-\infty}(\pi)$ the space of continuous <u>conjugate-linear</u> functionals on $H^\infty(\pi)$. The inclusion $H^\infty(\pi) \subset H(\pi)$ and the canonical isomorphism between a Hilbert space and its anti-dual then provide an inclusion of $H(\pi)$ into $H^{-\infty}(\pi)$:

$$H^\infty(\pi) \subset H(\pi) \subset H^{-\infty}(\pi) .$$

Let $T \mapsto T^*$ be the canonical involution on $U(\underline{g})$ such that $X^* = -X$ for $x \in \underline{g}$. By taking adjoints, we obtain a representation $X \mapsto \pi_{-\infty}(X)$ of $\underline{g}$ on $H^{-\infty}(\pi)$. The unitarity of π implies that $\pi_\infty(X)$ is skew-symmetric, and hence

$$(\pi_\infty(T)u,v) = (u, \pi_{-\infty}(T^*)v) ,$$

if $u \in H^\infty(\pi)$, $v \in H^{-\infty}(\pi)$, and $T \in U(\underline{g})$.

In the case $\pi = \rho$ is the right regular representation of G , the space $H^\infty(\rho)$ consists of all C^∞ functions φ on G such that $T\varphi \in L^2(G)$ for all left-invariant differential operators T on G . The space $H^{-\infty}(\rho)$ consists of all distributions on G of the form $\Sigma\, T_i f_i$, where $f_i \in L^2(G)$ and T_i are left-invariant differential operators on G (finite sum). For the representations π^λ of the Heisenberg group, $H^\infty(\pi^\lambda) = S(\mathbb{R}^n)$, the Schwartz space of rapidly-decreasing C^∞ functions, and $H^{-\infty}(\pi^\lambda)$ is the space of tempered distributions on $\mathbb{R}^n$.

Let $\varphi \in H^\infty(\rho)$. Then the Fourier transform $\hat{\varphi}$ is a <u>smoothing operator</u> :

$$(*) \qquad \hat{\varphi}(\lambda) : H(\pi^\lambda) \to H^\infty(\pi^\lambda)$$

(a.e. $[d\lambda]$). Furthermore, for every $T \in U(\underline{g})$, the Fourier transform of $\rho(T)\varphi$ is the operator-valued function $\lambda \mapsto \pi_\infty^\lambda(T)\, \hat{\varphi}(\lambda)$, and by the Plancherel formula,

$$(**) \qquad \int_{\mathbb{R}} \|\pi_\infty^\lambda(T)\, \hat{\varphi}(\lambda)\|_{HS}^2 \, d\mu(\lambda) < \infty .$$

Conversely, any measurable operator field $\hat{\varphi}$ which satisfies (*) and (**) for all $T \in U(\underline{g})$ is the Fourier transform of a function $\varphi \in H^\infty(\rho)$.

To obtain the Szegö kernel for $H^2(D)$, we shall need the (non-commutative) Fourier inversion formula for G . If A is a bounded operator on a Hilbert space, denote its <u>absolute value</u> by

$$|A| = (A^*A)^{1/2} .$$

Then $|A|$ is a non-negative self-adjoint operator, and A is <u>nuclear</u> if $|A|^{1/2}$ is Hilbert-Schmidt, i.e..

$$\Sigma(|A|e_n, e_n) < \infty ,$$

if $\{e_n\}$ is an orthonormal basis. In this case

$$tr(A) = \Sigma(Ae_n, e_n)$$

is defined independently of the choice of basis.

Define $L^1(\hat{G})$ to be the space of all measurable, operator-valued functions $\lambda \mapsto A(\lambda)$ from $\mathbb{R}$ to $L^2(\mathbb{R}^n)$ such that

$$\|A\|_1 = \int_{\mathbb{R}} tr(|A(\lambda)|)\, d\mu(\lambda) < \infty .$$

This is a Banach space in the norm $\|A\|_1$ (identifying null functions, as always). Given $A \in L^1(\hat{G})$, we define the <u>inverse Fourier transform</u> of A to be the function φ on G given by

$$(***) \qquad \varphi(g) = \int_{\mathbb{R}} tr(\pi^\lambda(g)\, A(\lambda))\, d\mu(\lambda) .$$

This defines a map from $L^1(\hat{G})$ into $C(G)$. If it happens that $\varphi \in L^2(G)$, then $\hat{\varphi}(\lambda) = A(\lambda)$ a.e. . For example, if we start with $\varphi \in H^\infty(\rho)$, then $\hat{\varphi} \in L^1(\hat{G})$, and (***) holds with $A(\lambda) = \hat{\varphi}(\lambda)$. (For proofs of the assertions of this section, see the notes and references at the end of the Chapter.)

2.2 <u>Tangential Chauchy-Riemann equations</u> Consider the domain

$$D = \{(z,w) : \operatorname{Im} w > \|z\|^2 , \quad z \in \mathbb{C}^n , \quad w \in \mathbb{C}\}$$

in $\mathbb{C}^{n+1}$, with boundary

$$M = \{(z,w) : \operatorname{Im}(w) = \|z\|^2\} ,$$

as in § II.4.4. The complex vector fields

$$\overline{L}_k = \frac{\partial}{\partial \overline{z}_k} - 2iz_k \frac{\partial}{\partial \overline{w}} \quad , \quad 1 \leq k \leq n$$

are tangent to M and span the anti-holomorphic tangent space at each point of M. Thus, if f is a function which is holomorphic on a neighborhood of M in $\mathbb{C}^{n+1}$, then

$$(\dagger) \qquad \overline{L}_k f = 0 \quad \text{on} \quad M , \quad 1 \leq k \leq n .$$

These are the "tangential Cauchy-Riemann equations."

Recall the lifting map W carrying functions on M to functions on G :

$$Wf(u) = f(e^{\lambda(u)}.o) ,$$

where λ is the Lie algebra homomorphism from $\underline{g}$ to vector fields on M defined in § II.4.4 . We know that W intertwines λ and the right regular representation ρ of $\underline{g}$. If $\{P_k, Q_k : 1 \leq k \leq n\}$ are the elements of $\underline{g}$ such that

$$\lambda(P_k) = X_k , \quad \lambda(Q_k) = Y_k$$

($X_k = \operatorname{Re}(L_k)$, $Y_k = \operatorname{Im}(L_k)$ as in § II.4.4) , then equations $(\dagger)$ become

$$(\dagger\dagger) \qquad \rho(A_k)\, Wf = 0 , \quad 1 \leq k \leq n ,$$

where

$$A_k = P_k - i\, Q_k .$$

Suppose now we start with a function $\varphi \in L^2(G)$, and consider the equations

$$(\#) \qquad \rho(A_k)\,\varphi = 0 \quad , \qquad 1 \leq k \leq n \ .$$

Here $\rho = \rho_{-\infty}$; i.e., we consider these as equations in the space $H^{-\infty}(\rho)$ introduced in the previous paragraph. By definition, this means that for all $\psi \in H^{\infty}(\rho)$, one has

$$(\#)' \qquad (\rho(A_k^*)\,\psi,\varphi) = 0 \quad , \quad 1 \leq k \leq n \ .$$

(The usual test-function space $C_c^{\infty}(G)$ is dense in $H^{\infty}(\rho)$, so this is the same as saying that $(\#)$ holds in the usual distribution sense.)

<u>Theorem</u> Suppose $\varphi \in L^2(G)$. Then φ satisfies the equations $(\#)$ $\Longleftrightarrow$ the Fourier transform $\hat{\varphi}$ satisfies the conditions

(i) $\hat{\varphi}(\lambda) = 0$ a.e. for $\lambda \leq 0$;

(ii) Range $\hat{\varphi}(\lambda) \subseteq V_\lambda$ a.e. for $\lambda > 0$,

where V_λ is the one-dimensional subspace of $L^2(\mathbb{R}^n)$ spanned by the function $\exp\left[-\frac{\lambda}{2}\|x\|^2\right]$ (the "vacuum state") .

<u>Proof</u> By the results of the previous section, the Fourier transform of $\rho(A_k^*)\psi$ is $\pi_\infty^\lambda(A_k^*)\,\hat{\psi}(\lambda)$. Hence

$$(\rho(A_k^*)\,\psi,\ \varphi) = \int_{\mathbb{R}} \operatorname{tr}(\pi_\infty^\lambda(A_k^*)\ \hat{\psi}\ (\lambda)\ \hat{\varphi}\ (\lambda)^*)d\mu(\lambda) \ .$$

We claim first that equations $(\#)$ are equivalent to

$$(\#\#) \qquad \text{Range } \hat{\varphi}(\lambda) \perp \text{Range } \pi_\infty^\lambda(A_k^*)$$

for $1 \leq k \leq n$ and a.e. λ .

It is immediate from the Plancherel formula that $(\#\#)$ implies $(\#)$. The converse follows from the existence of "sufficiently many" operators of the form $\hat{\psi}(\lambda)$. Specifically, if $f \in L^2(\mathbb{R}^n)$ and $h \in C_c(\mathbb{R})$, then the operator-valued function

$$\lambda \mapsto h(\lambda) \quad g \otimes f$$

is the Fourier transform of some $\psi \in H^{\infty}(\rho)$ (Here $g \otimes f$ is the operator of rank one on $L^2(\mathbb{R}^n)$ acting by $v \mapsto (v,f)g$.) This follows from the characterisation of $H^{\infty}(\rho)$ given in the previous paragraph, since the elements of $U(\underline{g})$ act in the representations π^{λ} as differential operators with polynomial coefficients. With this choice of ψ, we have

$$\pi_{\infty}^{\lambda}(A_k^*)\ \hat{\psi}(\lambda)\ \hat{\varphi}(\lambda)^* = h(\lambda)(\pi_{\infty}^{\lambda}(A_k^*)g) \otimes (\hat{\varphi}(\lambda)f) .$$

Hence by the Plancherel formula, (#) implies that

$$\int_{\mathbb{R}} h(\lambda)(\pi_{\infty}^{\lambda}(A_k^*)g,\ \hat{\varphi}(\lambda)f)\ d\mu(\lambda) = 0$$

for all such f, g, h (Recall that $tr(g \otimes f) = (g,f)$) . By the arbitrariness of h, this is equivalent to the conditions

$$(\pi_{\infty}^{\lambda}(A_k^*)g,\ \hat{\varphi}(\lambda)f) = 0 \qquad \text{a.e. } (\lambda)$$

for $1 \leq k \leq n$ and all $f \in L^2(\mathbb{R}^n)$, $g \in S(\mathbb{R}^n)$, which is precisely (##).

To investigate (##), we need the explicit form of the operators $\pi_{\infty}^{\lambda}(A_k^*)$. Going back to the formulas for π^{λ} , we calculate that

$$\pi_{\infty}^{\lambda}(P_k) = \frac{\partial}{\partial x_k} , \quad \pi_{\infty}^{\lambda}(Q_k) = i\lambda x_k ,$$

acting on $S(\mathbb{R}^n)$, where $\{x_k\}$ are the coordinate functions on $\mathbb{R}^n$. Since $A_k^* = -P_k - iQ_k$, this gives the formula

$$\pi_{\infty}^{\lambda}(A_k^*) = -\frac{\partial}{\partial x_k} + \lambda x_k .$$

For the operator $A_k = P_k - iQ_k$, we have

$$\pi_{\infty}^{\lambda}(A_k) = \frac{\partial}{\partial x_k} + \lambda x_k .$$

Case 1: If $\lambda < o$, the function $\exp(\frac{\lambda}{2}\|x\|^2)$ is in $S(\mathbb{R}^n)$, and is annihilated by $\pi_\infty^\lambda(A_k^*)$. Using the commutation relations

$$\left[\frac{\partial}{\partial x_k} + \lambda x_k, \ -\frac{\partial}{\partial x_k} + \lambda x_k\right] = 2\lambda\, I \quad ,$$

one verifies inductively that the subspace

$$W_\lambda = \bigcup_k \text{Range } \pi_\infty^\lambda(A_k^*)$$

contains the functions $p(x)\exp(\frac{\lambda}{2}\|x\|^2)$, where p is an arbitrary polynomial. This set of functions is dense in $L^2(\mathbb{R}^n)$, as is well-known, so that (##) $\Rightarrow$ $\hat{\varphi}(\lambda) = o$ when $\lambda < o$.

Case 2: If $\lambda > o$, the function $v_\lambda(x) = \exp(-\frac{\lambda}{2}\|x\|^2)$ is in $S(\mathbb{R}^n)$, and is annihilated by $\pi_\infty^\lambda(A_k)$. Hence it is orthogonal to W_λ. On the other hand, the same argument as in Case 1 shows that W_λ contains the functions $p(x)\, v_\lambda(x)$, where p is any polynomial such that $p(o) = o$. Hence $W_\lambda^\perp$ is one-dimensional and is spanned by the function v_λ in this case. This shows that (##) is equivalent to conditions (i) and (ii) of the Theorem, and completes the proof.

Corollary The space of functions $\varphi \in L^2(G)$ which satisfy

$$(\#) \qquad \rho(A_k)\,\varphi = o \ , \qquad 1 \le k \le n \quad ,$$

is a closed subspace of $L^2(G)$. The orthogonal projection P onto this subspace commutes with left translation. If $\psi \in L^2(G)$, then the Fourier transform of $P\psi$ is $P_\lambda\, \hat{\psi}(\lambda)$, where

$$P_\lambda = \begin{cases} 0 \ , & \text{if } \lambda < o \\ c(\lambda)\, v_\lambda \otimes v_\lambda \ , & \lambda > o \end{cases} \quad .$$

Here v_λ is the function $\exp[-\frac{\lambda}{2}\|x\|^2]$, and $c(\lambda) = \|v_\lambda\|^{-2} = (\lambda/\pi)^{n/2}$.

Proof: This follows immediately from the Plancherel theorem and the theorem just proved.

Definition Let $H_b^2(G)$ be the closed subspace of $L^2(G)$ consisting of all functions $\varphi \in L^2(G)$ which satisfy (#).

Remark Define $L^2(M)$ by transporting the Haar measure on G to a measure on M via the map $u \mapsto e^{\lambda(u)} \cdot 0$. If $H_b^2(M)$ is the subspace of functions in $L^2(M)$ which satisfy the tangential Cauchy-Riemann equations (†) (in the distribution sense), then

$$H_b^2(G) = W\, H_b^2(M) ,$$

where W is the lifting map from functions on M to functions on G, as in § II.4.4. (In this notation, b = boundary.)

2.3 Projection onto $H_b^2(G)$ as a singular integral operator Let $P : L^2(G) \to H_b^2(G)$ be the projection operator in Corollary 2.2. In this section we want to use the Fourier inversion formula on G (§2.1) to show that P is a linear combination of the identity operator and a singular integral operator of the type studied in Chapter III.

To illustrate the method, we first consider the analogous problem for the classical Paley-Wiener space $H_+^2(\mathbb{R})$, consisting of functions $f \in L^2(\mathbb{R})$ whose Fourier transform $\hat{f}(\lambda)$ vanishes for $\lambda \le 0$. If $P_+ : L^2(\mathbb{R}) \to H_+^2(\mathbb{R})$ is the orthogonal projection, then for $\varphi \in C_c^\infty(\mathbb{R})$ we can write, by the classical Fourier inversion formula,

$$P_+ \varphi(0) = \frac{1}{2\pi} \int_0^\infty \hat{\varphi}(\xi)\, d\xi$$

$$= \lim_{\substack{\varepsilon \to 0 \\ \varepsilon > 0}} \frac{1}{2\pi} \int_{\mathbb{R}} \left\{ \int_0^\infty e^{-\xi(\varepsilon + ix)}\, d\xi \right\} \varphi(x)\, dx$$

$$= \lim_{\substack{\varepsilon\to 0\\ \varepsilon>0}} \frac{1}{2\pi} \int_{\mathbb{R}} \frac{\varphi(x)}{\varepsilon+ix}\, dx \quad .$$

To write this limit as a singular integral, we observe that for any fixed $R > o$,

$$\lim_{\substack{\varepsilon\to 0\\ \varepsilon>0}} \frac{1}{2\pi} \int_{-R}^{R} \frac{dx}{\varepsilon+ix} = \frac{1}{\pi} \int_{0}^{\infty} \frac{dx}{x^2+1} = \frac{1}{2} \quad ,$$

independent of R . (Write the integral as an integral over $[o,R]$ and make the change of variable $x \to \varepsilon x$.) Since φ has compact support, we thus can write

$$P_+\varphi(o) = \lim_{\substack{\varepsilon\to 0\\ \varepsilon>0}} \frac{1}{2\pi} \int \frac{\varphi(x)-\varphi(o)}{\varepsilon+ix}\, dx + \frac{1}{2}\,\varphi(o)$$

$$= \frac{1}{2\pi i} \int \frac{\varphi(x)-\varphi(o)}{x}\, dx + \frac{1}{2}\,\varphi(o) \quad .$$

But the function x^{-1} has mean-value zero, so this last equation can also be written as

$$P_+\varphi(o) = \frac{1}{2\pi i} \lim_{\varepsilon\to 0} \int_{|x| > \varepsilon} \frac{\varphi(x)}{x}\, dx + \frac{1}{2}\,\varphi(o) \quad .$$

Finally, using the translation-invariance of P_+ , we conclude that

$$P_+ = \frac{1}{2}\,[I-iA] \quad ,$$

where A is the classical Hilbert transform (cf.§ 1.5).

Coming back to the Heisenberg group G and the projection P onto $H_b^2(G)$, we first observe that by the (non-commutative) Fourier inversion formula and Corollary 2.2,

$$P\varphi(e) = \int_0^\infty \operatorname{tr}(P_\lambda\, \hat{\varphi}(\lambda))\, d\mu(\lambda) \quad ,$$

if $\varphi \in C_c^\infty(G)$. Here $d\mu(\lambda) = c_n\, |\lambda|^n\, d\lambda$ is the Plancherel measure for G , and

$$P_\lambda \hat{\varphi}(\lambda) = c(\lambda)(v_\lambda \otimes v_\lambda)\hat{\varphi}(\lambda)$$

$$= c(\lambda)\, v_\lambda \otimes \hat{\varphi}(\lambda)^* v_\lambda \quad ,$$

where $v_\lambda(x) = \exp\left[-\frac{\lambda}{2}\|x\|^2\right]$ and $c(\lambda) = (\lambda/\pi)^{n/2}$. Hence

$$\mathrm{tr}\,(P_\lambda \hat{\varphi}(\lambda)) = c(\lambda)(\hat{\varphi}(\lambda)\, v_\lambda, v_\lambda)$$

$$= c(\lambda) \int_G \varphi(\lambda)(\pi^\lambda(g^{-1})\, v_\lambda, v_\lambda)\, dg \; .$$

Next, from the explicit form of the representations π^λ in § 2.1, a routine calculation shows that

$$(\pi^\lambda(g)\, v_\lambda, v_\lambda) = c(\lambda)^{-1} \exp\left[i\lambda\zeta - \frac{\lambda}{4}\left(\|\xi\|^2 + \|\eta\|^2\right)\right] ,$$

where the coordinates of g are (ξ,η,ζ) , as in § 2.1. Thus we can write $P\varphi$ as the iterated integral (non-absolutely convergent)

$$(*) \qquad P\varphi(e) = \int_0^\infty \left(\int_G e^{-\lambda\tau(g)}\, \varphi(g)\, dg\right) d\mu(\lambda) \quad ,$$

where

$$\tau(g) = \frac{1}{4}\left(\|\xi\|^2 + \|\eta\|^2\right) - i\zeta$$

if $g = (\xi,\eta,\zeta)$.

To interchange the order of integration in $(*)$, we introduce a convergence factor $\exp(-\varepsilon\lambda)$, $\varepsilon > 0$, so that

$$P\varphi(e) = \lim_{\substack{\varepsilon\to 0\\ \varepsilon>0}} \int_G \left(\int_0^\infty e^{-\lambda(\varepsilon+\tau(g))}\, d\mu(\lambda)\right) dg \; .$$

Using the formula for the Plancherel measure to evaluate the inner integral, we obtain the formula

$$(**) \qquad P\varphi(e) = \lim_{\substack{\varepsilon\to 0\\ \varepsilon>0}} c \int_G \frac{\varphi(g)}{(\varepsilon+\tau(g))^{n+1}}\, dg \quad ,$$

where $\quad c = n!(2\pi)^{-n-1}$, and $\varphi \in C_c^\infty(G)$.

It remains to rewrite (**) as a principal-value integral of the type considered in Chapter III. For dilations on G we take the one-parameter group of automorphisms whose action in canonical coordinates is

$$\delta_t(\xi,\eta,\zeta) = (t\xi, t\eta, t^2\zeta) \quad ,$$

when $t > o$. The function τ introduced above is then homogeneous of degree 2 :

$$\tau(\delta_t g) = t^2\tau(g) \ .$$

We define a homogeneous norm on G by setting

$$|g| = |\tau(g)|^{1/2}$$

$$= [\zeta^2 + 4^{-2}(\|\xi\|^2 + \|\eta\|^2)^2]^{1/4} \ .$$

This norm is smooth and symmetric. (Recall that g^{-1} has coordinates $(-\xi,-\eta,-\zeta)$.) The homogeneous dimension of G is $Q = 2n+2$, and the function

$$K(g) = \tau(g)^{-n-1}$$

is homogeneous of degree $-Q$. Obviously K is C^∞ on $G \sim \{e\}$. Also

$$K(g^{-1})^* = K(g) \quad ,$$

so K is Hermitian symmetric.

<u>Lemma</u>. K has mean-value zero.

<u>Proof</u> Introduce "cylindrical" coordinates on G by setting $\rho = (1/4)(\|\xi\|^2 + \|\eta\|^2)$. The change of measure is then

$$d\xi \, d\eta \, d\zeta = c \, \rho^{n-1} \, d\omega \, d\rho \, d\zeta \quad ,$$

where $d\omega$ is the measure on the unit sphere in $\mathbb{R}^{2n}$, and c is a positive constant. In these coordinates, we have

$$\int_{a\leq|g|\leq b} K(g)\,dg = c \iint_E (\rho-i\zeta)^{-n-1}\,\rho^{n-1}\,d\rho\,d\zeta \quad ,$$

where $E \subset \mathbb{R}^2$ is the half annulus

$$a^4 \leq \zeta^2 + \rho^2 \leq b^4 \quad , \quad \rho \geq o \quad .$$

Changing to polar coordinates in the ρ-ζ plane, we get the integral

$$2c \int_a^b \{\int_o^\pi e^{i(n+1)\theta} (\cos\theta)^{n-1}\, d\theta\} \frac{dr}{r} \; .$$

Writing $\cos\theta = \frac{1}{2}(e^{i\theta} + e^{-i\theta})$ and using the binomial expansion, we find that the integrand is a sum of even harmonics $\exp(2ik\theta)$, $1 \leq k \leq n$, and hence it integrates to zero over $[o,\pi]$, Q.E.D.

We can now state the main result of this section. Let us denote by K the singular convolution operator

$$K\,\varphi(x) = \text{P.V.} \int_G \varphi(xy)\,K(y)\,dy \; .$$

By the lemma just proved and the results of Chapter III, we know that K extends to a bounded operator on $L^2(G)$.

<u>Theorem</u> The projection P onto $H_b^2(G)$ is a linear combination of K and the identity operator.

<u>Proof</u> Since P and K commute with left translations, it is enough to express $P\,\varphi(e)$ in terms of $K\,\varphi(e)$ and $\varphi(e)$, for $\varphi \in C_c^\infty(G)$. By using the same cylindrical coordinates as in the lemma, we find that

$$\lim_{\substack{\varepsilon\to o\\ \varepsilon>0}} \int_{|g|\leq R} \frac{dg}{(\varepsilon+\tau(g))^{n+1}} \equiv a$$

exists and is independent of R . Since $(\varphi-\varphi(e))K$ is an integrable function if $\varphi \in C_c^\infty(G)$, we thus obtain from (**) the equation

$$P\varphi(e) = c \int_G (\varphi(g) - \varphi(e))\, K(g)\, dg + a\, \varphi(e)\ .$$

But K has mean-value zero, so the first term on the right side of this equation is $cK\varphi(e)$. This shows that

$$P = cK + aI\ ,$$

finishing the proof.

Remark For the three-dimensional Heisenberg group $(n=1)$, the operator K is the intertwining operator studied in § 1.5 associated with the representation π_2 of $SU(2,1)$.

2.4 Szegö kernel for $H^2(D)$ Let

$$D = \{(z,w) : z \in \mathbb{C}^n\ ,\ w \in \mathbb{C}\ ,\ \operatorname{Im}(w) > \|z\|^2\}$$

be the Siegel domain of type II introduced in § II. 4.4 . If $(z,w) \in D$, we can write

$$(z,w) = (z,\ w_o + it)\ ,$$

where $\operatorname{Im} w_o = \|z\|^2$ and $t > o$. Thus the point $(z,w_o) \in M$. Given a function f on D , we define the function f_t on M by

$$f_t(z,w_o) = f(z,w_o + it)\ , \qquad t > o\ .$$

The Hardy space $H^2(D)$ is then defined to be the space of all holomorphic functions f on D such that

$$(*) \qquad \sup_{t>o} \int_M |f_t(m)|^2\, dm < \infty\ .$$

Recall that the measure dm is the image of Haar measure on the Heisenberg group (Equivalently, we can parametrize M by $\mathbb{R}^n \times \mathbb{R}^n \times \mathbb{R}$ via the map $(z,w) \mapsto (\mathrm{Re}\, z, \mathrm{Im}\, z, \mathrm{Re}\, w)$ and use Lebesgue measure in the parameters.)

In this section we will use the Fourier analysis on the Heisenberg group to show that: (i) the boundary values of functions in $H^2(D)$ comprise the space $H_b^2(M)$ already studied; (ii) with norm given by the left side of $(*)$, $H^2(D)$ is a Hilbert space, and the mapping from f to its boundary function is an isomorphism onto $H_b^2(M)$; (iii) the function f can be recovered from its boundary values by an integral formula.

To illustrate the method, we first consider the classical case of the half-plane $\{\mathrm{Im}\, z > o\} \subset \mathbb{C}$. Starting with $\varphi \in H_+^2(\mathbb{R})$, we obtain a holomorphic function f in $\mathrm{Im}\, z > o$ by the Fourier inversion formula:

$$f(z) = \frac{1}{2\pi} \int_0^\infty e^{i\xi z}\, \hat{\varphi}(\xi)\, d\xi \quad , \qquad \mathrm{Im}\, z > o \quad .$$

(Recall that $\hat{\varphi}(\xi) = o$ for $\xi \le o$.) If we define $f_t(x) = f(x+it)$, then f_t is the inverse Fourier transform of the function $\exp(-t\xi)\, \hat{\varphi}(\xi)$, so that by the Plancherel theorem,

$$\sup_{t>o} \|f_t\|_{L_2(\mathbb{R})} = \|\varphi\|_{L_2(\mathbb{R})} .$$

In particular, the set of functions $\{f_t\}_{t>o}$ is bounded in $L_2(\mathbb{R})$, and $f_t \to \varphi$ in L_2 as $t \to o$. Conversely, given a holomorphic function f in the upper half-plane with the property that the set $\{f_t\}_{t>o}$ is bounded in $L^2(\mathbb{R})$, we can use a weak-compactness argument to obtain a boundary function $\varphi \in H_+^2(\mathbb{R})$, such that $f_t \to \varphi$ as $t \to o$. Finally, to represent f in terms of φ instead of $\hat{\varphi}$, we invert the order of integration (no convergence factor is needed now, since $\mathrm{Im}\, z > o$) to get the formula

$$f(z) = \frac{1}{2\pi} \int_{-\infty}^{\infty} \{\int_0^\infty e^{i\xi(z-x)}\, d\xi\}\, \varphi(x)\, dx$$

$$= \frac{1}{2\pi i} \int_{-\infty}^{\infty} \frac{\varphi(x)}{x-z}\, dx \quad .$$

The function $(2\pi i)^{-1}(x-z)^{-1}$ is the Szegö kernel in this case (the reproducing kernel expressing $f(z)$ in terms of the boundary values of f).

We now return to the space $H^2(D)$, and carry out a similar analysis, replacing Fourier analysis on $\mathbb{R}$ by (non-commutative) Fourier analysis on the Heisenberg group G. Recall that from § II. 4.4 the map from G to M is given in coordinates by

$$(**)\qquad \begin{aligned} w(g) &= \zeta + \frac{i}{4}\left(\|\xi\|^2 + \|\eta\|^2\right) \\ z_j(g) &= \frac{1}{2}(\xi_j - i\eta_j) \quad , \end{aligned}$$

when g has canonical coordinates $(\xi,\eta,\zeta) \in \mathbb{R}^n \times \mathbb{R} \times \mathbb{R}$. We shall write

$$z(g) = (z_1(g),\ldots,z_n(g)) \in \mathbb{C}^n \ ,$$

and if $z,z' \in \mathbb{C}^n$, we set

$$z\cdot z' = \sum_{j=1}^{n} z_j z'_j \quad .$$

Note that as a real C^∞ manifold, D is isomorphic to $G \times \mathbb{R}_+$ via the map

$$g,t \mapsto (z(g) \quad , \quad w(g) + it) \quad ,$$

where $g \in G$ and $t > o$.

The first step in the analysis is to show that by analytic continuation of the Fourier inversion formula for G, we can synthesize functions in $H^2(D)$, starting from the Fourier transforms of functions in $H^2_b(M)$. For this, we need the following consequence of Theorem 2.2:

Lemma The Fourier transform of the space $H^2_b(G)$ consists of all operator-valued functions $\lambda \mapsto v_\lambda \otimes w_\lambda$, where $v_\lambda, w_\lambda \in L_2(\mathbb{R}^n)$ satisfy

(i) $v_\lambda(x) = \exp\left[-\frac{\lambda}{2}\|x\|^2\right]$

(ii) $w_\lambda = o$ if $\lambda \leq o$

(iii) the function $\lambda, x \mapsto w_\lambda(x)$ is measurable on $\mathbb{R}_+ \times \mathbb{R}^n$,

and

$$\int_0^\infty \int_{\mathbb{R}^n} |w_\lambda(x)|^2 \ \lambda^{n/2} \ dx \ d\lambda \ < \infty \ .$$

(Here $v \otimes w$ denotes the operator on $L_2(\mathbb{R}^n)$ given by $(v \otimes w)(u) = (u,w)\, v$.)

If $\varphi \in H_b^2(G)$ and $\hat{\varphi}(\lambda) = v_\lambda \otimes w_\lambda$, then the map $\varphi \mapsto \overline{w}_\lambda$ defines a Hilbert space isomorphism from $H_b^2(G)$ onto $L^2(\mathbb{R}_+ \times \mathbb{R}^n \ ; \ \nu)$, where $d\nu = c\, \lambda^{n/2}\, d\lambda\, dx$ ($d\lambda$ = Lebesgue measure on $\mathbb{R}$, dx = Lebesgue measure on $\mathbb{R}^n$, c = constant).

<u>Proof of Lemma</u> Conditions (i) and (ii) are direct consequence of Theorem 2.2. To verify condition (iii) , we observe that the Hilbert-Schmidt norm of the operator $v_\lambda \otimes w_\lambda$ is $\|v_\lambda\| \, \|w_\lambda\|$. Hence by the Plancherel formula,

$$\int_G |\varphi(g)|^2 \, dg = c_n \int_0^\infty \|v_\lambda\|^2 \ \|w_\lambda\|^2 \ \lambda^n \ d\lambda$$

$$= c \int_0^\infty \int_{\mathbb{R}^n} |w_\lambda(x)|^2 \ \lambda^{n/2} \ dx \ d\lambda \ ,$$

since $\|v_\lambda\|^2 = (\pi/\lambda)^{n/2}$. (Here $c = \pi^{n/2} c_n$, where $c_n = (2\pi)^{-n-1}$.) Together with the Plancherel theorem, this proves the Lemma.

The main result of this section is the following ($d\mu = c_n |\lambda|^n d\lambda$ is the Plancherel measure for G) :

<u>Theorem</u> Suppose $\varphi \in H_b^2(G)$. Then for every $t > o$, the operator-valued function

$$F_t(\lambda) = e^{-\lambda t} \ \hat{\varphi}(\lambda)$$

is in $L^1(\hat{G})$. If the function f is defined on the Siegel domain D by

$$(\#) \qquad f(z(g), w(g) + it) = \int_0^\infty tr(\pi^\lambda(g)\, F_t(\lambda)) \ d\mu(\lambda) \ ,$$

then $\qquad f \in H^2(D) \quad$ and

$$\sup_{t>o} \|f_t\|_{L_2(M)} = \|\varphi\|_{L_2(G)} .$$

Furthermore, $Wf_t \to \varphi$ in $L_2(G)$ as $t \searrow o$.

<u>Proof</u> By the Lemma just proved, we can write $\hat{\varphi}(\lambda) = v_\lambda \otimes w_\lambda$, and hence

$$tr(\pi^\lambda(g)\ \hat{\varphi}(\lambda)) = (\pi^\lambda(g)\ v_\lambda, w_\lambda)$$

(inner product in $L_2(\mathbb{R}^n)$) . Using the explicit form of the representation π^λ , we calculate that

$$(\pi^\lambda(g)\ v_\lambda, w_\lambda) = e^{i\lambda w - \lambda z \cdot z} \int_{\mathbb{R}^n} e^{-\frac{\lambda}{2}[x \cdot x + 4z \cdot x]}\ \overline{w_\lambda(x)}dx ,$$

where $w = w(g)$ and $z = z(g)$ are defined by (**) . This makes it evident that the integrand on the right side of (#) is a holomorphic function on D .

The trace norm of the operator $F_t(\lambda)$ is given by

$$\|F_t\|_1 = e^{-\lambda t}\ \|v_\lambda\|\ \|w_\lambda\| = e^{-\lambda t}\ \|\hat{\varphi}(\lambda)\|_2^2$$

It follows by the Plancherel theorem that (#) is absolutely convergent, uniformly in $g \in G$ and uniformly for t in compact subsets of (o,∞) . Hence f is the limit of holomorphic functions,uniformly on compact subsets of D , so f is holomorphic. Clearly Wf_t is the L_2 function on G whose Fourier transform is $e^{-\lambda t}\ \hat{\varphi}(\lambda)$. Hence by the Plancherel theorem $\{f_t\}$ is a bounded subset of $L_2(M)$, and $Wf_t \to \varphi$ in $L_2(G)$ as $t \searrow o$, Q.E.D.

<u>Remark</u> By the Theorem, we can pass from functions in $H_b^2(G)$ to functions in $H^2(D)$, via the Fourier inversion formula. To see that we obtain all of $H^2(D)$ in this way, we can argue as follows: If $f \in H^2(D)$, then for each $t > o$, f_t satisfies the tangential Cauchy-Riemann equations (since f is holomorphic). Hence $\{Wf_t\}$ is a bounded subset of $H_b^2(G)$. Since $H_b^2(G)$ is a closed subspace of $L_2(G)$, there is an element $\varphi \in H_b^2(G)$ and a subsequence t_k such that

$$Wf_{t_k} \to \varphi \quad \text{weakly} \ .$$

On the other hand, since f is holomorphic, the Cauchy-Riemann equations give the relation

$$\frac{\partial}{\partial t} Wf_t = i \frac{\partial}{\partial \zeta} Wf_t$$

Taking Fourier transforms, we conclude that

$$\frac{\partial}{\partial t} (Wf_t)^\wedge = -\lambda \ (Wf_t)^\wedge$$

and hence

$$(Wf_t)^\wedge(\lambda) = e^{-\lambda t} \ \hat{\varphi}(\lambda) \ .$$

The proof of the Theorem then shows that f is obtained from φ by (#).

We conclude our study of the space $H^2(D)$ by rewriting formula (#) in terms of the boundary values of f, eliminating the Fourier transform. Given $\varphi \in H^2_b(G)$, we write $\hat{\varphi}(\lambda) = v_\lambda \otimes w_\lambda$, by the Lemma. Then

$$w_\lambda = c(\lambda) \ \hat{\varphi}(\lambda)^* \ v_\lambda \ , \qquad \text{where} \quad c_\lambda = \|v_\lambda\|^{-2} \ .$$

Hence we can write

$$\begin{aligned} tr(\pi^\lambda(g) \ \hat{\varphi}(\lambda)) &= c(\lambda)(\pi^\lambda(g)v_\lambda, \ \hat{\varphi}(\lambda)^* v_\lambda) \ , \\ &= c(\lambda)(\hat{\varphi}(\lambda)\pi^\lambda(g)v_\lambda, \ v_\lambda) \\ &= c(\lambda) \int_G (\pi^\lambda(\gamma^{-1}g)v_\lambda, \ v_\lambda) \ \varphi(\gamma) \ d\gamma \ . \end{aligned}$$

(The last step is a-priori true if φ also is in $L_1(G)$.) But from § 2.3 we have the formula

$$c(\lambda)(\pi^\lambda(g)v_\lambda, \ v_\lambda) = e^{i\lambda w(g)} \ .$$

Hence formula (#) of the Theorem can be written as

$$f(z(g),\ w(g)+it) = c_n \int_0^\infty\int_G \exp[-\lambda t+i\lambda w(\gamma^{-1}g)]\ \varphi(\gamma)\lambda^n\ d\gamma\ d\lambda$$

$$= n!c_n \int_G [t-iw(\gamma^{-1}g)]^{-n-1}\ \varphi(\gamma)\ d\gamma\ .$$

(This last integral is easily seen to be absolutely convergent, for any $t > o$ and $\varphi \in L^2(G)$, so by Fubini and dominated convergence this justifies the earlier steps.)

To write this formula in terms of the holomorphic coordinates on the ambient space $\mathbb{C}^{n+1}$, we calculate that

$$w(\gamma^{-1}g) = w(g) - \overline{w(\gamma)} - 2i\ z(g) \cdot \overline{z(\gamma)}\ .$$

Hence if we denote the boundary values of f on M by f also, then we have the integral formula

$$f(z,w) = d_n \int_M [w-\overline{w}'-2i\ z\cdot\overline{z}']^{-n-1}\ f(z',w')\ dm\ ,$$

where (z',w') are the coordinates on M, and $d_n = (-1)^{n+1}\ n!\ (2\pi i)^{-n-1}$. Thus the <u>Szegö kernel</u> for the space $H^2(D)$ is the function

$$S(p,p') = d_n\ [w-\overline{w}'-2i\ z\cdot\overline{z}']^{-n-1}\ ,$$

where $p = (z,w) \in D$ and $p' = (z',w') \in M$.

§ 3 Hypoelliptic differential operators

3.1 Fundamental solutions for homogeneous hypoelliptic operators Let D be a differential operator (with C^∞ coefficients) on a C^∞ manifold M. Recall that D is said to be hypoelliptic if every distribution solution f to the equation $Df = g$ satisfies

$$\text{Sing Supp}(f) = \text{Sing Supp}(g) .$$

Here Sing Supp(f) denotes the singular support of f, i.e. the complement of the open set on which f is a C^∞ function.

From the point of view of analysis on nilpotent groups, one of the most interesting examples is an operator of the form

$$D = X_o + \sum_{j=1}^{n} X_j^2 ,$$

where $X_o, X_1, \ldots X_n$ are real vector fields on a manifold M. It is a fundamental theorem of L. Hörmander that such an operator is hypoelliptic provided the Lie algebra generated by $X_o, X_1, \ldots, X_n$ spans the tangent space at each point of M. This is precisely the infinitesimal transitivity hypothesis that was the starting point of our constructions in Chapter II.

The first step in our analysis of hypoelliptic operators associated with transitive Lie algebras of vector fields will be to study the corresponding operators on a graded nilpotent group. In fact, it is only the graded vector space structure and the hypothesis of hypoellipticity that we need at first.

Assume V is a real, graded vector space with dilations $\{\delta_t\}$, as in § I.1.1. An operator D on $C^\infty(V)$ will be called homogeneous of degree α if

$$D(\varphi\circ\delta_t) = t^\alpha(D\varphi) \circ \delta_t$$

for all $\varphi \in C^\infty(V)$.

Theorem Suppose D is a C^∞ differential operator on V which is homogeneous of degree α , with $0 < \alpha < Q$ = homogeneous dimension of V . Assume that D and its transpose D^t are both hypoelliptic. Then there exists a unique distribution K on V which is homogeneous of degree $\alpha - Q$ and satisfies $DK = \delta$. (δ = delta function at 0) .

Remark Hypoellipticity and the condition $\alpha > o$ imply that K is defined by a function $k(x)$, which is C^∞ away from 0 and homogeneous of degree $\alpha - Q$. (Note that k is locally integrable, and that no distribution supported at $\{o\}$ can be homogeneous of degree $> -Q$.)

Proof of Theorem (Sketch) The hypoellipticity of D and its transpose, together with some general functional analysis, imply that

1) There exists a distribution K_0 , defined on some neighborhood $U = \{|x| < a\}$ of zero, which satisfies $DK_0 = \delta$ on U ;

2) On the subspace $N = \{\varphi \in C^\infty(U) : D\varphi = 0\}$, the distribution topology and the C^∞ function topology coincide.

The idea of the proof is to take the distribution K_0 , which a priori has no particular homogeneity properties, and to construct $h_0 \in N$ such that the distribution $K = K_0 - H_0$ is homogeneous of degree $\alpha - Q$ (H_0 being the distribution $h_0(x)\,dx$ on U) . Of course, K also satisfies $DK = \delta$.

Since the set U is invariant under dilations $\{\delta_t : 0 < t \leq 1\}$, we can define distributions H_t , $0 < t \leq 1$ by

$$< H_t, \varphi > = < K_0, \varphi > - t^{-\alpha} < K_0, \varphi \circ \delta_{1/t} > ,$$

for $\varphi \in C_c^\infty(U)$. If K_0 were homogeneous of degree $\alpha - Q$, then H_t would be zero. In any event, the assumption that D is homogeneous of degree α, and the fact that the delta function is homogeneous of degree $-Q$, imply that

$$DH_t = \delta - \delta = 0 .$$

Hence by hypoellipticity of D, H_t is of the form $h_t(x)\,dx$, with $h_t \in C^\infty(U)$.

We want to show that $\lim\limits_{t\to o} h_t$ exists. Away from 0, K_o is given as $k_o(x)\,dx$, where $k_o \in C^\infty(U\sim\{o\})$. The formula above for H_t can be written in pointwise terms as

$$h_t(x) = k_o(x) - t^{Q-\alpha}\, k_o(\delta_t x) ,$$

for $x \in U \sim \{o\}$ and $o < t \leq 1$. Hence

$$h_s(x) - h_r(x) = r^{Q-\alpha}\, k_o(\delta_r x) - s^{Q-\alpha}\, k_o(\delta_s x)$$

$$= r^{Q-\alpha}\, h_{s/r}\,(\delta_r x) ,$$

if $o < s \leq r \leq 1$. (Since $h_t \in C^\infty(U)$, this formula also holds at $x = o$.) Taking $s = r^2$, we obtain the recursive relation

$$h_{r^2}(x) = r^{Q-\alpha}\, h_r(\delta_r x) + h_r(x) .$$

Iterating this relation, we find that

$$(*) \qquad h_{r_n}(x) = \sum_{k=o}^{2^n-1} r^{k(Q-\alpha)}\, h_r(\delta_{r^k} x) ,$$

where $r_n = r^{2^n}$.

The mapping $t \mapsto H_t$ is evidently continuous from $(0,1)$ into $D'(U)$, and hence by property 2) above the map $t \mapsto h_t$ is continuous from $(0,1)$ into $C^\infty(U)$ (relative to the topology of uniform convergence of functions and their derivatives on compact subsets of U). In particular, $\{h_t : \frac{1}{4} \leq t \leq \frac{1}{2}\}$ is a compact subset of $C^\infty(U)$. Hence if $\varepsilon < a$, then

Then K_1 is a kernel of type m, and we let A_1 be the corresponding operator.

By the results of § 3.2, we can extend formula (III) of § III.5.3 as follows: If φ is a function on X of the form $\varphi(x) = f(\theta(y,x))$, then

$$(\lambda(P)\varphi)(x) = (dR(P)f)(\theta(y,x)) + (T_y f)(\theta(y,x)) \quad ,$$

where T_y is a differential operator on Ω of order $\leq m - 1$ at $v = 0$. Taking $f = k_1$, we see that the first term of this formula gives a delta function, and the second term gives a kernel of type 1. Multiplying k_1 by the truncation functions a and b only contributes another kernel of type 1 in the calculation of $\lambda(P)_x K_1(x,y)$. The same is true for the action of the lower order term R on A_1 by virtue of Theorem 2 of § III.5.3. This proves the first equation. The operator Δ^t also satisfies the hypotheses, by Lemma 3.2. This gives the second equation, Q.E.D.

Corollary Suppose the Lie algebra V is generated by its elements of degree one. Assume that $f \in L^p(X)$ satisfies

$$\Delta f = g \quad ,$$

where $g \in S^p_k(X)$. Then $a f \in S^p_{k+m}(X)$ for every function $a \in C^\infty_c(X)$.

Proof Given a, we construct operators A of type m and T of type 1, such that

$$A \Delta = a I + T \quad .$$

Then by Corollary III.5.4, we have $A g \in S^p_{k+m}$ and $T f \in S^p_1$. Hence $a f \in S^p_1$. Iterating this argument $k + m$ times proves the Corollary.

§ 3.4 Local regularity We continue in the same context as § 3.3, but now we only assume that

$$\lambda : V \to L(M)$$

is a <u>transitive</u> partial homomorphism (surjective onto the tangent space at each point of M .) Let the spaces $DO(\lambda)_m$ of differential operators of λ-degree $\leq m$ be defined as in § III.5.3 , Definition 3 .

<u>Definition</u> If $1 < p < \infty$ and m is a non-negative integer, then $S^p_{m,loc}(M,\lambda)$ consists of all distributions f on M such that

$$DO(\lambda)_m \, f \subset L^p_{loc}(M)$$

(i.e. if $v_i \in V$ are of degree n_i , and $n_1+\cdots+n_k \leq m$, then the distribution derivative

$$\lambda(v_1) \cdots \lambda(v_k) \, f \in L^p_{loc}(M) \text{ .)}$$

<u>Remark</u> It is clear from the formula for the transpose of a vector field that this definition is independent of the choice of measure μ on M .

In order to apply the results of § 3.3 to the present context, we need to know how the spaces $S^p_{m,loc}$ behave under liftings of transitive partial homomorphisms.

Fix a point $x_o \in M$, and recall the lifting map

$$W : C^\infty(M) \to C^\infty(\Omega) \quad ,$$

where Ω is a sufficiently small open set around 0 in V . For any function on M , let

$$Wf(v) = f(e^{\lambda(v)}x_o) \quad .$$

Let $\Lambda : V \to L(\Omega)$ be a lifting of λ as constructed in Chapter II . For $f \in L^1_{loc}(M)$, define $\tilde{W}f$ to be the distribution $Wf(v)\,dv$ on Ω .

Lemma 1. There is an open set M_o around x_o, a surjective linear map

$$P : C_c^\infty(\Omega) \to C_c^\infty(M_o) \quad ,$$

and a linear map

$$\gamma : V \to C^\infty(\Omega) \quad ,$$

such that

$$(1) \qquad < \tilde{W}f, \varphi > = < f, P\varphi >$$

$$(2) \qquad \lambda(v) P\varphi = P(\Lambda(v)\varphi + \gamma(v)\varphi) \quad .$$

Here $f \in L^1_{loc}(M)$, $\varphi \in C_c^\infty(\Omega)$ and the pairing in (1) is given by integration over Ω and M, respectively.

Remark Recall that in the case that λ is an exact homomorphism, then W is the map lifting functions from a homogeneous space to the group. The operator P is "integration over the cosets" in this case, and is a standard tool in harmonic analysis.

Proof of Lemma 1. We use the description of the range of W in local coordinates given in § II.2.4 : There are coordinates $\{x_j\}$ on Ω so that Wf is a function of $x' = (x_1,\dots,x_m)$ only $(m = \dim M)$. Set $x'' = (x_{m+1},\dots,x_n)$, and view x' as giving coordinates on M_o.

If $f \in L^1_{loc}(M)$ and $\varphi \in C_c^\infty(\Omega)$, then with this identification we can write

$$< \tilde{W}f, \varphi > = \iint f(x')\, \varphi(x';x'')\, J(x';x'')\, dx'\, dx'' \quad .$$

Here dx' and dx'' denote Lebesgue measure in the respective coordinates, and $J > 0$ is a suitable change of measure factor. Thus if we define

$$P\varphi(x') = \int \varphi(x';x'')\, J(x';x'')\, dx'' \quad ,$$

then condition (1) is obviously satisfied. The surjectivity of P is evident on taking functions of the form $\varphi_1(x')\,\varphi_2(x'')$.

To verify (2), we express the vector fields $\lambda(v)$ and $\Lambda(v)$ in these coordinates in the form

$$\lambda(v) = a(x')\,\partial/\partial x'$$

$$\Lambda(v) = a(x')\,\partial/\partial x' + b(x',x'')\,\partial/\partial x''$$

(such an expression is equivalent to the relation $\Lambda(v)\,W = W\,\lambda(v)$) . Integrating by parts in the integral for $P(\Lambda(v)\varphi)$, we obtain relation (2), finishing the proof.

If f is a distribution on M_o , we now define $\tilde{W}f$ as a distribution on Ω by formula (1). Then $\tilde{W}$ maps the space of distributions on M_o injectively into the space of distributions on Ω , since P is surjective. Because of the density factors in the measures on M and Ω , the intertwining relation between $\lambda(v)$ and $\Lambda(v)$ now becomes

$$\tilde{W}\,\lambda(v) = (\Lambda(v) + a_v)\,\tilde{W} \quad , \tag{3}$$

where $a_v \in C^\infty(M_o)$.

Lemma 2 Let f be a distribution on M_o . Then

$$f \in S^p_{m,loc}(M_o,\lambda) \iff \tilde{W} f \in S^p_{m,loc}(\Omega,\Lambda) \quad .$$

Proof The Lemma is clearly true when $m = o$, by definition of $\tilde{W}f$. Suppose f and $\lambda(v)\,f$ are both locally in L^p . Then there is a function $g \in L^p_{loc}(M_o)$ such that

$$< f,\ \lambda(v)\,P\varphi > \;=\; < g,\ P\varphi >$$

for all $\varphi \in C^\infty_c(\Omega)$. Using relation (3) , we find that

$$< \tilde{W} f, \Lambda(v)\varphi > = < \tilde{W} g - a_v, \varphi > \quad ,$$

which shows that $\Lambda(v)\tilde{W} f \in L^p_{loc}(\Omega)$. Iterating this argument gives the implication $\Rightarrow$.

Conversely, if $f \in L^p_{loc}(M_o)$ and $\Lambda(v)\tilde{W} f \in L^p_{loc}(\Omega)$, then (3) shows that there is a function $G \in L^p_{loc}(\Omega)$ such that

$$< f, \lambda(v)\, P\varphi > = < G,\varphi >$$

for all $\varphi \in C^\infty_c(\Omega)$. Using the formula for P in local coordinates (x',x'') and taking φ of the form $\varphi_1(x')\, \varphi_2(x'')$, we find by Hölder's inequality that

$$< f, \lambda(v)\psi > = < g,\psi > \ , \quad \text{if} \quad \psi \in C^\infty_c(M_o) \quad ,$$

where $g \in L^p_{loc}(M_o)$. By induction this finishes the proof.

We turn finally to the main result of this section. Assume now that V is the free , r-step nilpotent Lie algebra on n generators, with its standard gradation (cf. Chapter II, § 1.2). In this case the "Lifting Theorem" of Chapter II asserts that there exists a lifting Λ of λ which is also a partial homomorphism from V to vector fields on Ω . This allows us to transfer the results of § 3.3, as follows (Q=homogeneous dimension of V):

Theorem Let Δ be a differential operator on M , such that Δ has a "principal part" $\lambda(P)$, where P is homogeneous of degree m , $0 < m < Q$. Assume that the left-invariant differential operators $dR(P)$ and $dR(P^t)$ are hypoelliptic. Then if $f \in L^p_{loc}(M)$ satisfies the equation $\Delta f=g$, where $g \in S^p_{k,loc}(M,\lambda)$, it follows that $f \in S^p_{k+m,loc}(M,\lambda)$. (Here $1 < p < \infty$.)

Proof The result is local, so we may work in a neighborhood of a fixed point x_o on which the Lifting Theorem can be applied. Since V is a free nilpotent Lie algebra, there exists a partial homomorphism Λ which is a lifting of λ. Thus Λ satisfies the hypotheses of § 3.3 .

By relation (3) of this section, we find that the distribution $\tilde{f} = \tilde{W}f$ satisfies an equation of the form

$$D\tilde{f} = \tilde{g} \quad ,$$

where $\tilde{g} = \tilde{W}g$, and D is an operator with principal part $\Lambda(P)$. By Lemma 2 and the Corollary to the Theorem of § 3.3, we conclude that

$$\tilde{f} \in S^p_{k+m,loc}(\Omega,\Lambda) \ .$$

Hence $f \in S^p_{k+m,loc}(M,\lambda)$, Q.E.D.

Example Let Δ be the operator considered in the example at the end of § 3.2, and assume that the vector fields $\{X_i : 1 \leq i \leq n\}$ and their commutators up to length r suffice to span the tangent space at each point of M. Assume also that either $\dim M > 2$, or else $\dim M = 2$ and $r > 1$. Then $n \geq 2$, and the homogeneous dimension Q of the free, r-step nilpotent Lie algebra on n generators satisfies $Q > 2$, so the theorem just proved applies to Δ. For example, if $f \in L^p_{loc}(M)$ is such that $\Delta f \in L^p_{loc}(M)$, then we can conclude that

$$X_i f \ , \quad X_i X_j f \in L^p_{loc}(M)$$

also, for $1 \leq i, j \leq n$. Note that this result involves no explicit mention of nilpotent groups or singular integral operators.

Comments and references for Chapter IV

§ 1 We have largely followed Knapp-Stein [1], to which we refer the reader for more examples and references. The literature on intertwining operators for semi-simple groups is extensive. For the basic analytic properties of the "intertwining integral", cf. Kunze-Stein [1] and Schiffmann [1]. For recent developments, we cite Johnson-Wallach [1], Helgason [2], Knapp-Stein [2], [3]. See also Stein's survey talk [1], Wallach [1], and Warner [1].

§ 2.1 The use of non-commutative Fourier analysis on the Heisenberg group to study the space $H^2(D)$ was first done by Ogden and Vâgi [1], who consider the general "Siegel domain of type II". For a survey of the unitary representation theory of nilpotent groups, cf. Moore [1]. The C^∞ regularity theory is treated; e.g. in Goodman [1], [2], [4], Poulsen [1], and Cartier [1], [2].

§ 2.2 The Theorem is due to Ogden-Vâgi [1]. Our proof, using the behaviour of C^∞ vectors under direct integral decomposition, is slightly different. The "tangential Cauchy-Riemann equations" can be written more intrinsically in terms of the operator $\bar{\partial}_b$; cf. Folland-Kohn [1].

§ 2.3 The Theorem is due to Korânyi-Vâgi [1]; cf. Korânyi-Vâgi-Welland [1].

§ 2.4 For the general theory of the Szegö kernel of a domain, cf. Gindikin [1] and Stein [2]. For the Szegö kernel in the case of Siegel domains, cf. Korânyi [1]. The construction of the Szegö kernel given here, using the Fourier analysis on the Heisenberg group, avoids the problem of proving *a priori* that functions in $H^2(D)$ have a boundary integral representation. For connections between Szegö kernels and representations of semi-simple groups, cf. Knapp [1].

§ 3.1 For the hypoellipticity of second-order operators, cf. Hörmander [1] and Kohn [1], [2]. The Theorem and its proof are taken from Folland [2]. The results from functional analysis cited in the proof can be found in Trèves [1], § 52. The heat equation, relative to the subelliptic Laplacian in examples 3 and 4, has been studied by Hulanicki [1], Folland [2], and Jørgensen [1]. The fundamental solution for the subelliptic Laplacian on the Heisenberg group was calculated by Folland [1]. Gaveau [1] has used stochastic integrals to calculate fundamental solutions on two-step nilpotent groups. Rockland [1] has shown that on the Heisenberg group, a homogeneous, left-invariant differential operator is hypoelliptic, provided its image in every non-trivial irreducible unitary representation has a bounded inverse. See Grušin [1], [2] for examples of hypoelliptic operators with polynomial coefficients.

§ 3.2 For the relations between tensor algebras, universal enveloping algebras, and left-invariant differential operators on a Lie group, cf. Helgason [1]. The study of operators whose principal part is a polynomial in the vector fields $\lambda(v)$ with variable coefficients has been treated, in special cases, by Rothschild-Stein [1].

§§ 3.3-3.4 These results are due to Rothschild-Stein [1], extending similar results of Folland-Stein [1]. The proofs given here are somewhat different, since we have again tried to emphasize the similarity with the case of functions on a homogeneous space, as in Chapter II.

For the classical elliptic regularity theory we refer to Bers-John-Schechter [1]. For applications of the regularity results here to the study of the $\bar{\partial}_b$ operator, we refer to Folland-Stein [1] and Rothschild-Stein [1]. For elliptic regularity in the context of unitary representation theory, cf. Goodman [4].

Appendix

Generalized Jonquières Groups

In Chapter I we restricted our attention to the group of automorphisms of the ring P of polynomial functions, and we embedded every simply-connected nilpotent Lie group as a subgroup of such a group. For geometric reasons it is desirable to consider a larger group, namely the group of automorphisms of the field of rational functions (the Cremona group). In this appendix we want to construct certain (finite dimensional) Lie subgroups of the Cremona group, extending the constructions of § I.1. In fact, the construction works over any field of characteristic zero. We will assume in these notes that the coefficient field is the complex numbers, however, since we have systematically avoided any mention of algebraic groups up to this point.

The first step in this analysis is to replace the one-parameter dilation group on the vector space V by an n-parameter dilation group $(n = \dim V)$. The generators of this group span a commutative subalgebra $\underline{h} \subset \mathrm{Der}(P)$, and the adjoint action of $\underline{h}$ on $\mathrm{Der}(P)$ is diagonalizable.

In § A.1 we study the vector fields with polynomial coefficients which are eigenvectors for $\mathrm{ad}(\underline{h})$. In § A.2, we construct a family of maximal finite-dimensional subalgebras of $\mathrm{Der}(P)$, each of which contains $\underline{h}$, and determine their structure in § A.3. (The nilpotent algebras studied in Chapter I occur as subalgebras of these maximal algebras.) In order to achieve maximality, we have to include vector fields which generate birational (but not everywhere-defined) transformations of V. In § A.4 we construct groups of birational transformations corresponding to these maximal subalgebras.

A.1 Root space decomposition of Der(P)

Let P be the algebra of polynomial functions on a finite-dimensional complex vector space V. Fix a basis $\{x_i\}$ for V and dual basis $\{\xi_i\}$ for V^*, $1 \le i \le d$. Thus the monomials $\{\xi^a : a \in \mathbb{N}^d\}$ are a basis for P, and the vector fields $\{\xi^a \partial_i : a \in \mathbb{N}^d, 1 \le i \le d\}$ are a basis for the Lie algebra $\underline{g} = \mathrm{Der}(P)$ $(\partial_i = D_{x_i})$.

Consider the commutative subalgebra

$$\underline{h} = \mathrm{span}\,\{\xi_i \partial_i \quad : \quad 1 \le i \le d\} .$$

The action of $\underline{h}$ on P is diagonalizable. Indeed, if we write $H_i = \xi_i \partial_i$, then

$$H_i\, \xi^a = a_i\, \xi^a .$$

Hence if we define linear functions μ_a, $a \in \mathbb{N}^d$, on $\underline{h}$ by $\mu_a(H_i) = a_i$, and set $M = \{\mu_a : a \in \mathbb{N}^d\}$, then

$$P = \Sigma\, H_\mu \qquad (\mu \in M)$$

where

$$H_\mu = \{f \in P : H f = \mu(H) f \quad (H \in \underline{h})\} .$$

Thus $\dim H_\mu = 1$, and H_μ has basis ξ^a if $\mu = \mu_a$.

The action of ad$\underline{h}$ on $\underline{g}$ can also be diagonalized. Indeed, we have the commutation relations

$$(1) \qquad [\xi^a \partial_i, \xi_j \partial_j] = (\delta_{ij} - a_j)\, \xi^a \partial_i$$

(δ_{ij} = Kronecker delta). Hence if we define the linear functions $\lambda_{a,i}$ on $\underline{h}$ by $\lambda_{a,i}(H_j) = \delta_{ij} - a_j$, and set $L = \{\lambda_{a,i} : 1 \le i \le d, a \in \mathbb{N}^d\}$, then

$$\underline{g} = \Sigma\, \underline{g}_\lambda \qquad (\lambda \in L) ,$$

where

$$\underline{g}_\lambda = \{X \in \underline{g} \ : \ [X,H] = \lambda(H)X\}$$

From the Jacobi identity and the formula $H(Xf) = XHf + [H,X]f$, we obtain the relations

$$(2) \qquad \begin{cases} H_\mu H_\nu \subseteq H_{\mu+\nu} \\ [\underline{g}_\kappa , \underline{g}_\lambda] \subseteq \underline{g}_{\kappa+\lambda} \\ \underline{g}_\lambda H_\mu \subseteq H_{\mu-\lambda} \end{cases}$$

(Define $H_\mu = 0$, $\underline{g}_\lambda = 0$ if $\mu \notin M$ or $\lambda \notin L$.)

It is evident that $\underline{h} = \underline{g}_0$ and that the normalizer of $\underline{h}$ in $\underline{g}$ is $\underline{h}$ itself, so that $\underline{h}$ is a Cartan subalgebra of $\underline{g}$. The root spaces $\underline{g}_\lambda$, $\lambda \neq 0$, are of two basic types:

<u>Proposition</u> If $\lambda \in L$, then one of the following situations occurs:

(I) For all i , $\lambda(H_i)$ is a non-positive integer. In this case $\dim \underline{g}_\lambda = d$ and $\underline{g}_\lambda = H_{-\lambda}\underline{h}$.

(II) There is an index i such that $\lambda(H_i) = 1$, while $\lambda(H_j)$ is a non-positive integer for all $i \neq j$. In this case $\dim \underline{g}_\lambda = 1$.

<u>Proof</u> It is clear from formula (1) that $\lambda \in L$ satisfies (I) or (II). In case (I) , there are d choices of $a \in \mathbb{N}^d$ and i so that $\lambda = \lambda_{a,i}$, while in case (II), there is a unique choice of a,i . The corresponding vector fields $\xi^a \partial_i$ are all linearly independent and span $\underline{g}_\lambda$, Q.E.D.

<u>Definition</u> (I) We will write $\lambda < 0$ if $\lambda \neq 0$ and $\lambda(H_i)$ is a non-positive integer for all i .

(II) We will call $\lambda \in L$ <u>elementary</u> if it satisfies condition (II) of the proposition.

<u>Remarks</u> The dichotomy described in the proposition is reflected in the geometric properties of the vector fields in $\underline{g}_\lambda$, as follows:

1. Suppose $\lambda = \lambda_{a,i}$ is elementary. Then $\underline{g}_\lambda$ has basis $X_\lambda = \xi^a \partial_i$. Furthermore, since $\lambda(H_i) = 1$, this forces $a_i = 0$. Hence the function ξ^a is independent of ξ_i , so that the flow generated by X_λ is

$$\begin{cases} \xi_i \to \xi_i + t\xi^a \\ \xi_j \to \xi_j \, , \quad j \neq i \end{cases}$$

We shall call this an elementary flow, and refer to the induced automorphisms of P as <u>elementary</u> automorphisms. (More generally, an automorphism of P defined by

$$\begin{cases} \xi_i \to c\xi_i + f(\xi_1, \ldots, \xi_{i-1}, \xi_{i+1}, \ldots, \xi_d) \\ \xi_j \to \xi_j \, , \quad j \neq i \, , \end{cases}$$

where f is a polynomial and $c \neq 0$, is called a <u>Jonquières</u> transformation.)

2. Suppose $\lambda = 0$. Then $\underline{g}_0 = \underline{h}$, and the vector field $\Sigma\, c_i H_i$ generates the flow

$$\xi_i \to \xi_i \exp(tc_i) \quad .$$

3. Suppose $\lambda < 0$ and $X \in \underline{g}_\lambda$, $X \neq 0$. Then X does <u>not</u> generate an automorphism of P . To see this, use the fact that $X = \xi^a H$, where $H \in \underline{g}_0$ and $a_i = -\lambda(H_i) \geq 0$. Then for any $b \in \mathbb{N}^d$, one calculates easily that

$$X^n (\xi^b) = c_n \, \xi^{na + b} \quad ,$$

where $c_n = \beta(H)(\beta(H) + \alpha(H))\cdots(\beta(H) + (n-1)\alpha(H))$. Here $\alpha, \beta \in M$ are the linear forms on $\underline{h}$ corresponding to $a, b \in \mathbb{N}^d$. But we can always pick b so that $-\beta(H) / \alpha(H) \notin \mathbb{N}$. This choice makes $c_n \neq 0$ for all n . Hence X does not act locally nilpotently on P , and the formal series $\Sigma(1/N!)X^n$ does not define an automorphism of P .

4. When $\lambda < 0$, it is possible for certain vector fields in $\underline{g}_\lambda$ to generate automorphisms of larger algebras of functions than the algebra P. The simplest example, perhaps, is obtained by taking $X = \xi^a H$ as above and requiring that H be orthogonal to α, i.e. $\alpha(H) = 0$. The calculation above then shows that

$$\Sigma(1/n!)X^n \xi^b = \xi^b \exp\left[\beta(H)\xi^a\right] .$$

Hence X generates an automorphism of the algebra of entire functions (= power series with infinite radius of convergence) in this case.

5. When the local flow defined by $X \in \underline{g}_\lambda$ is rational, then X generates a one-parameter group of automorphisms of the field R of rational functions. For example, let $X = \xi_1 H$, where

$$H = \Sigma\, c_i H_i \quad , \quad c_1 = 1 \quad .$$

The local flow generated by X is obtained by solving the system of differential equations

$$\begin{cases} x_k'(t) = c_k x_1(t) x_k(t) \\ x_k(0) = \xi_k \end{cases}$$

One finds that

$$x_k(t) = \xi_k\,(1-t\xi_1)^{-c_k} \quad .$$

Hence if $c_k \in \mathbb{Z}$, then the map $\xi_k \to x_k$ is rational for each t, with rational inverse

$$\xi_k = x_k\,(1+tx_1)^{-c_k}$$

Hence $\exp tX$ is defined for all t as an automorphism of R.

A.2 Maximal finite-dimensional subalgebras.

Having studied various one-parameter groups of automorphisms generated by vector fields in the root spaces $\underline{g}_\lambda$, we turn to the problem of finding finite-dimensional Lie groups of automorphisms of the polynomial algebra, or of the algebra of rational functions. As a first step, we shall look for finite dimensional subalgebras $\underline{m}$ of the Lie algebra $\underline{g}$ which are maximal in the following two senses:

(1) $\underline{m} \supset \underline{h}$, so that rank $(\underline{m}) = \dim V$.

(2) $\underline{m}$ is not contained in any larger finite-dimensional subalgebra of $\underline{g}$.

We will construct an infinite family of such maximal algebras as follows: Choose an element $H \in \underline{h}$ which is positive integral in the sense that

$$H = \Sigma\, n_i H_i \qquad (H_i = \xi_i \partial/\partial\xi_i) \quad ,$$

where the n_i are positive integers and $n_1 = 1$. By permuting the basis H_i , we arrange that

$$1 = n_1 = \cdots = n_q < n_{q+1} \leq \cdots \leq n_d = r \quad ,$$

where $1 \leq q \leq d$ is admitted. Then H generates the one-parameter group of linear transformations

$$\delta_t\, x_i = t^{n_i}\, x_i \quad , \qquad (t \neq 0) \; .$$

Let $V = V_1 \oplus \cdots \oplus V_r$ be the associated decomposition of V , where

$$V_k = \text{span}\,\{x_i : n_i = k\} \quad .$$

Let $V_k^* = \text{span}\,\{\xi_i : n_i = k\}$ be the dual space. The element H also defines a hyperplane in $\underline{h}^*$. We shall construct a finite-dimensional algebra $\underline{m}$ by taking all the root spaces on or above this hyperplane, together with q one-dimensional root spaces lying one step below the hyperplane.

In more detail, define

$$\begin{cases} \underline{n}_k = \Sigma \{\underline{g}_\lambda : < \lambda, H > = k\} \\ \underline{n}_- = \{\xi H : \xi \in V_1^*\} \end{cases} .$$

<u>Theorem</u> $\underline{m} = \underline{n}_- + \sum_{k \geq o} \underline{n}_k$ is a maximal finite-dimensional subalgebra of Der(P) .

<u>Proof</u> Since $[\underline{g}_\lambda, \underline{g}_\mu] \subseteq \underline{g}_{\lambda+\mu}$, it is clear that

$$[\underline{n}_k, \underline{n}_j] \subseteq \underline{n}_{k+j} .$$

Furthermore, $H(\xi_i) = n_i$, so that $\underline{n}_- \subseteq \underline{n}_{-1}$.

Hence

$$[\underline{n}_-, \underline{n}_k] \subseteq \underline{n}_{k-1} .$$

To show that $\underline{m}$ is a Lie algebra, we must prove that

$$[\underline{n}_-, \underline{n}_o] \subseteq \underline{n}_- .$$

We note first that $[\underline{h}, \underline{n}_-] \subseteq \underline{n}_-$, since $\underline{n}_-$ is the sum of the root spaces $(\xi_j H)$ with $n_j = 1$. If $\underline{g}_\lambda \subset \underline{n}_0$, $\lambda \neq 0$, then $\underline{g}_\lambda$ has basis $\xi^a \partial_i$, where $\Sigma\, n_k a_k = n_i$ and $a_i = 0$. Since n_k is increasing in k , this forces $a_k = 0$ for all k with $n_k > n_i$. We calculate the commutator to be

$$[\xi^a \partial_i, \xi_j H] = \delta_{ij}\, \xi^a H .$$

This vanishes unless $i = j$, and in this case $\xi^a = \xi_k$ for some k between 1 and q , since $n_j = 1$.

The fact that $H(\xi_j) = \xi_j$, $1 \leq j \leq q$, gives

$$[\underline{n}_-, \underline{n}_-] = 0 .$$

This shows that $\underline{m}$ is a Lie subalgebra of Der(P).

Each space $\underline{n}_k$ is certainly finite-dimensional. Indeed, $\underline{n}_k$ for $k > o$ is spanned by the elementary vector fields $\xi^a \partial_i$ with $a_i = 0$ and

$$\Sigma\, n_j a_j = n_i - k \ .$$

For each (k, i) there are only a finite number of $a \in \mathbb{N}^d$ satisfying this equation. Furthermore, since $n_i \leq r$, we have no solutions if $k > r$. This shows that

$$\underline{n}_k = 0 \ , \qquad k > r \ ,$$

and completes the proof that $\underline{m}$ is a finite-dimensional algebra containing $\underline{h}$.

It remains to show that there is no finite-dimensional subalgebra of $\underline{g}$ which properly contains $\underline{m}$. Since any such subalgebra would also contain the Cartan subalgebra $\underline{h}$, it would be the sum of its intersections with the root spaces $\underline{g}_\lambda$. Hence to prove that $\underline{m}$ is maximal, it suffices to show that if $Y \in \underline{g}_\lambda$ but $Y \notin \underline{m}$, then X and Y do not generate a finite-dimensional Lie subalgebra of $\underline{g}$, where $X = \xi_1 H$. Since $Y \notin \underline{m}$, we have $< \lambda, H > \leq -1$. Also $X \in \underline{g}_\mu$, where $< \mu, H > = -1$. If X and Y generated a finite-dimensional Lie algebra, we would thus have to have $(\mathrm{ad}X)^n (Y) = 0$ for some n, since this element lies in $\underline{g}_{\lambda+n\mu}$.

Since $\underline{g}_{\lambda+n\mu} \not\subset \underline{m}$ for any $n \geq 1$, we need only prove the following lemma to complete the proof of the theorem:

<u>Lemma</u> If $Y \in \underline{g}_\lambda$ but $Y \notin \underline{m}$, then $[X,Y] \neq 0$, where $X = \xi_1 H$.

<u>Proof</u>. There are two cases to consider. If $Y = \xi^a \partial_i$ is an elementary vector field, then $Y \notin \underline{m}$ means that $\Sigma\, n_k a_k = n_i + m$ where $m > 0$. We calculate the commutator

$$[Y,X] = \delta_{i1}\, \xi^a H - m\, \xi_1 \xi^a \partial_i \ ,$$

which is clearly not zero. If Y is not an elementary vector field, then $Y = \xi^a Z$, where $Z = \Sigma\, c_k H_k$ and $a \neq 0$. In this case

$$[Y,X] = \xi^a (c_1 H - bZ) \ ,$$

where $b = \Sigma a_k n_k$. This commutator can vanish only if $c_1 H = bZ$. In particular, this would imply that $c_1 \neq 0$ (since $b \neq 0$) , and hence $b = 1$. But this is only possible when there exists a $j \leq q$ such that $a_j = 1$ and $a_k = 0$, $k \neq j$. In this case we would have $Y \in \underline{m}$, proving the lemma.

Remark The element H generates the one-parameter group of dilations $x_i \to t^{n_i} x_i$ $(t \neq 0)$ on $V_{\mathbb{R}}$ $(= \text{span}_{\mathbb{R}} \{x_i\})$. The subalgebra

$$\underline{n} = \sum_{k \geq 1} \underline{n}_k \subset \underline{m}$$

is the complexification of the algebra associated with this dilation group which was studied in § 1.3 of Chapter I.

Example Let $n = 2$, and write $x = \xi_1$, $y = \xi_2$. Then $\underline{h}$ has basis $H_1 = x\partial_x$ and $H_2 = y\partial_y$. Identify the roots $\lambda \in L$ with the integral point $(\lambda(H_1), \lambda(H_2))$ in the plane. Then L consists of all points

$$\{(1,-n), (-m,1), (-m,-n) : m,n \in \mathbb{N}\} .$$

The corresponding root spaces are spanned by the following vector fields:

$$(1,-n) : y^n \partial_x$$

$$(-m,1) : x^m \partial_y$$

$$(-m,-n) : x^{m+1} y^n \partial_x , \quad x^m y^{n+1} \partial_y .$$

Fix an integer $n \geq 1$, set $H = H_1 + nH_2 = x\partial_x + ny\partial_y$, and let $\underline{m}$ be the algebra defined by H as in the theorem. Then the roots of $\underline{h}$ on $\underline{m}$ are as follows (cf. Fig 1):

	Roots		Root Vectors	
$n = 1$:	$(-1,0)$,	$(1,0)$	xH ,	∂_x
	$(-1,1)$,	$(1,-1)$	$x\partial_y$,	$y\partial_x$
	$(0,-1)$,	$(0,1)$	yH ,	∂_y

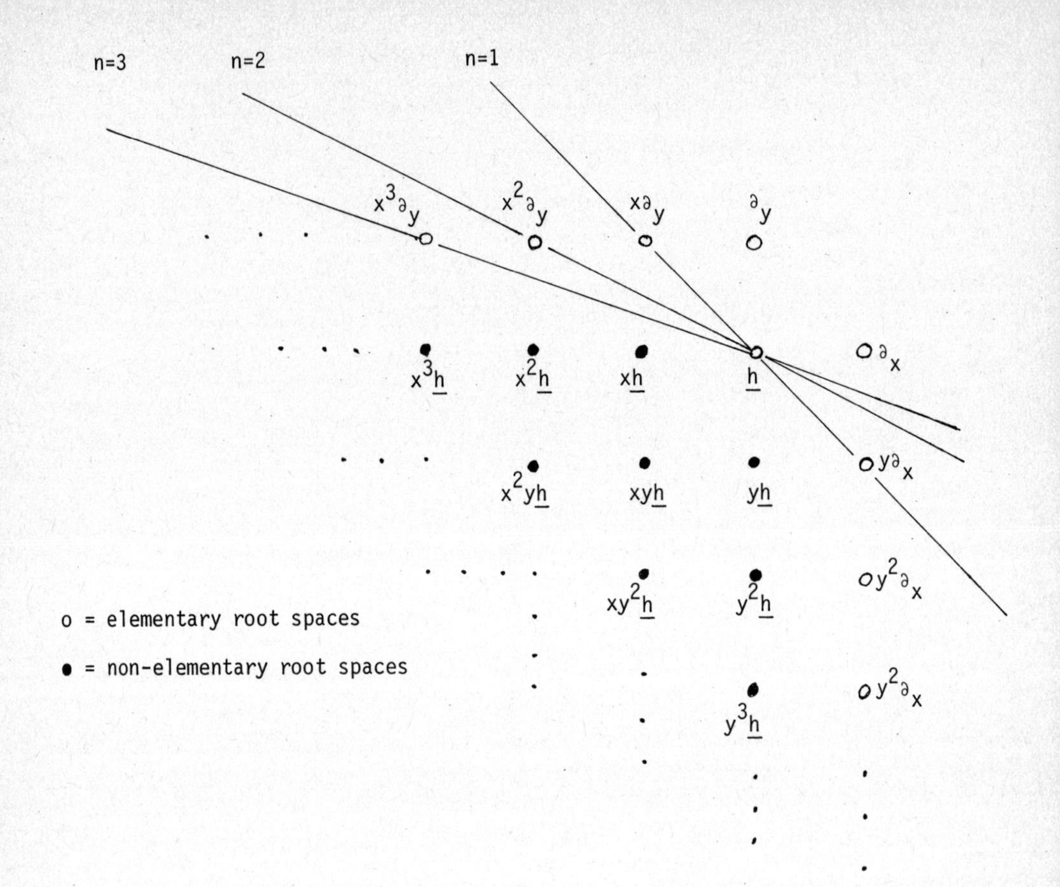

Fig 1 Root spaces for Der(P)

(P = polynomial algebra in two variables)

$$n > 1: \begin{cases} (-1,0) \ , \quad (1,0) & xH \ , \quad \partial_x \\ (-k,1) \ , \quad 0 \le k \le n & x^k \partial_y \ , \quad 0 \le k \le n \end{cases}$$

Note that in the case $n = 1$, the negative of every root is again a root. One verifies that $\underline{m} \simeq s\ell(3)$ in this case (cf. § A.3).

When $n > 1$, $\underline{m}$ is the semi-direct sum of

$$\underline{r} = (xH) \oplus \underline{h} \oplus (\partial_x)$$

and

$$\underline{u} = \text{span } \{x^k \partial_y \ : \ 0 \le k \le n\} \ .$$

Here $\underline{r}$ is the sum of $\underline{h}$ and the root spaces $\underline{m}_\lambda$ for those λ such that $-\lambda$ is also a root. One verifies that

$$\underline{r} \simeq s\ell(2) \oplus g\ell(1) \ ,$$

with the $s\ell(2)$ part spanned by xH, ∂_x, and $H+H_1$, while the $g\ell(1)$ factor is spanned by H_2. The ideal $\underline{u}$ is the sum of the root spaces $\underline{m}_\lambda$ such that $-\lambda$ is not a root. $\underline{u}$ is commutative, and one calculates that $ad(r)$ acts on $\underline{u}$ via the $(n+1)$ - dimensional irreducible representation of $s\ell(2)$, while $ad(H_2) = -1$ on $\underline{u}$ (cf. § A.3).

A.3 <u>Structure of $\underline{m}$</u> Let $\underline{m}$ be the maximal finite-dimensional subalgebra of Der P in § A.2, defined by a choice of $H \in \underline{h}$, where

$$H = \Sigma\, n_i H_i \quad , \quad 1 = n_1 \le n_2 \le \cdots \le n_d = r \ .$$

This choice of H determines a decomposition

$$V = V_1 \oplus \cdots \oplus V_r \quad ,$$

$(V_k = \text{span } \{x_i : n_i = k\})$, and determines a gradation $\{H_n\}$ of the polynomial functions on V.

We have

$$H_n = \Sigma H_\lambda \qquad (< H,\lambda > = n) .$$

Since $\underline{h} \subset \underline{m}$, we have a root-space decomposition

$$\underline{m} = \underline{h} + \Sigma \underline{m}_\lambda \ , \qquad (\lambda \in \Delta)$$

where $\underline{m}_\lambda = \underline{m} \cap \underline{g}_\lambda$, and $\Delta = \Delta(H)$ is a finite set of non-zero linear forms on $\underline{h}$. By the construction of $\underline{m}$, it is evident that

$$\Delta = \{\lambda \in L : < H,\lambda > \geq 0\} \cup \{-\lambda_j : 1 \leq j \leq q\} ,$$

where $\lambda_j(H_i) = \delta_{ij}$ and $q = \dim V_1$. In particular,

$$\underline{m}_\lambda = \underline{g}_\lambda \qquad \text{if} \quad < H,\lambda > \ \geq \ 0 \ .$$

We want to separate $\underline{m}$ into a semi-direct sum of a reductive and a nilpotent Lie algebra. For this it is natural to examine the set

$$\Delta_r = \Delta \cap (-\Delta) .$$

Since $< H,\lambda > \geq -1$, every root in Δ_r must satisfy $< H,\lambda > = 0$ or ± 1 .

Consider first the case $< H,\lambda > = 0$:

Lemma 1. Set $\underline{r}_o = \underline{h} + \Sigma \{\underline{g}_\lambda : \lambda \in \Delta_r$ and $< H,\lambda > = 0\}$. Then

$$\underline{r}_o \cong \sum_{k=1}^{r} \oplus \, g\ell \, (V_k) ,$$

where the isomorphism is defined by the action of $g\ell \, (V_k)$ on P induced by the decomposition $V = \Sigma V_k$.

Proof If $< H,\lambda > = 0$, $\lambda \neq 0$, then $\underline{g}_\lambda$ has basis $\xi^a \partial_i$, where $\Sigma n_j a_j = n_i$ and $a_i = 0$. If $-\lambda \in \Delta$ also, this forces ξ^a to be linear, and hence

$$\xi^a = \xi_j \quad (n_i = n_j) .$$

The vector fields $\{\xi_i\partial_j : n_i = n_j = k\}$ span the Lie algebra of $g\ell(V_k)$, and $\underline{r}_o$ is the sum of these subalgebras, Q.E.D.

Next, we observe that the commutative subalgebra

$$\underline{n}_+ = \mathcal{D}_1 = \{D_x : x \in V_1\}$$

is isomorphic to V_1 as an $\underline{h}$ module. The contragredient module is isomorphic to

$$\underline{n}_- = V_1^* H \quad .$$

<u>Lemma</u> 2 Set $\underline{r} = \underline{h} + \Sigma\, \underline{m}_\lambda \qquad (\lambda \in \Delta_r)$.

Then

$$\underline{r} = \underline{n}_- \oplus \underline{r}_o \oplus \underline{n}_+ \quad .$$

One has the commutation relations

$$\begin{cases} [\underline{n}_-, \underline{n}_-] = [\underline{n}_+, \underline{n}_+] = 0 \\ [\underline{r}_o, \underline{n}_\pm] \subseteq \underline{n}_\pm \\ [\underline{n}_-, \underline{n}_+] \subseteq \underline{r}_o \end{cases}$$

<u>Proof</u>. It is clear from the remarks above that $\underline{n}_-$ is the sum of the roots spaces of the $\lambda \in \Delta_r$ with $< H,\lambda > = -1$. Since $\underline{n}_+$ is the contragredient $\underline{h}$-module, it must therefore be the sum of the root spaces of the $\lambda \in \Delta_r$ with $< H,\lambda > = +1$. This exhausts the possibilities for $\lambda \in \Delta_r$. Hence $\underline{r} = \underline{n}_- \oplus \underline{r}_o \oplus \underline{n}_+$

Using the description of $\underline{r}_o$ in lemma 1, we easily verify the commutation relations. In particular, we note that $[D_x, H] = D_x$ if $x \in V_1$. Hence for $x \in V_1$ and $\xi \in V_1^*$, we have

$$[D_x, \xi H] = < x,\xi > H + \xi D_x \in \underline{h} \quad .$$

<u>Remark</u> One easily verifies that $(q = \dim V_1)$:

$$\underline{r} \cong s\ell(q+1) \oplus g\ell(V_2) \oplus \cdots \oplus g\ell(V_r) ,$$

and hence $\underline{r}$ is reductive. (View $s\ell(q+1)$ as the matrices $\begin{pmatrix} A & x \\ \xi & -trA \end{pmatrix}$, where $A \in g\ell(V_1)$, $x \in V_1$, $\xi \in V_1^*$. Such a matrix corresponds to the vector field $D_x - \xi H - \Sigma\, a_{ij}\, \xi_i\, \partial_j - tr(A)H$, if (a_{ij}) is the matrix of A on V_1 .)

To complete our analysis of the Lie algebra $\underline{m}$, we set

$$\Delta_u = \Delta \sim \Delta_r ,$$

and define

$$\underline{u} = \sum_{\lambda \in \Delta_u} \underline{g}_\lambda .$$

Since $< H,\lambda > \geq 0$ for all $\lambda \in \Delta_u$, we have

$$\underline{u} = \sum_{k \geq o} \underline{u}_k ,$$

where

$$\underline{u}_k = \Sigma \{\underline{g}_\lambda : \lambda \in \Delta_u \text{ and } < H,\lambda > = k\}$$

It is clear from lemma 2 that $\underline{u}_k = \underline{n}_k$ if $k \geq 2$. Furthermore, by § 1.3 and the description of $\underline{n}_+$, we see that

$$\underline{u}_1 = \{X \in \underline{n}_1 : X \text{ vanishes at } 0\}$$

Thus it remains to describe $\underline{u}_o$.

<u>Lemma</u> 3 Define $H_o^o = 0$ and

$$H_m^o = \Sigma\, H_j H_k \quad (j+k=m,\ j<m,\ k<m) .$$

Then

$$u_o = \{X \in Der(P) : X\, H_m \subseteq H_m^o \text{ for all } m\}$$

<u>Proof</u> Denote by $\underline{w}$ the set of vector fields X such that $X\, H_m \subseteq H_m^o$ for all m . $\underline{w}$ is obviously a Lie algebra. Since

$$H_k = \{f \in P : Hf = kf\} ,$$

it is clear that H_k and H_k^o are invariant under $\underline{h}$. Hence $\underline{w}$ is invariant under $ad\underline{h}$, and is thus the sum of root spaces.

Let $X = \xi^a \partial_i$ be a root vector in $\underline{w}$. Since X is a vector field of degree zero, relative to the gradation $\{H_n\}$, we have

$$\Sigma\, n_j a_j = n_i \quad .$$

We claim that $a_j = 0$ for all j such that $n_i = n_j$. Indeed if not, then we would have $\xi^a = \xi_j$, and thus $X\xi_i = \xi_j \notin H^o_{n_i}$, a contradiction.

Conversely, if $X = \xi^a \partial_i$ is a vector field such that

$$\Sigma\, n_j a_j = n_i \ , \quad a_j = 0 \ \text{ if } \ n_i = n_j \quad ,$$

then we claim that $X \in \underline{w}$. Indeed, it is clear that in this case

$$\xi^a \in H^o_{n_i}$$

so that

$$X\, H_m \subseteq H^o_{n_i}\, H_{m-n_i} \quad .$$

But we see from the definition that

$$H^o_n\, H_m \subseteq H^o_{m+n} \quad .$$

This proves that $X \in \underline{w}$.

Finally, if λ is any root of $\underline{h}$ on $\underline{g}$ such that $< H,\lambda > = 0$, and $-\lambda$ is not a root, then by the proof of lemma 1 $\underline{g}_\lambda$ has basis $\xi^a \partial_i$, where

$$\Sigma\, n_i a_j = n_i \ , \quad a_j = 0 \ \text{ for } \ n_i = n_j \quad .$$

This proves that $\underline{u}_o = \underline{w}$, Q.E.D.

We can now combine all the pieces and obtain the structure of $\underline{m}$:

Theorem Let Δ be the non-zero roots of $\underline{h}$ on $\underline{m}$. Divide Δ into $\Delta_r = \Delta \cap (-\Delta)$ and $\Delta_u = \Delta \sim \Delta_r$, and set

$$\underline{r} = \underline{h} + \Sigma\, \underline{m}_\lambda \qquad (\lambda \in \Delta_r)$$

$$\underline{u} = \Sigma\, \underline{m}_\lambda \qquad (\lambda \in \Delta_u) \ .$$

Then $\underline{m} = r \oplus u$ (vector space direct sum), and

(i) $\underline{r}$ is a reductive subalgebra of $\underline{m}$

(ii) $\underline{u}$ is a nilpotent ideal of $\underline{m}$

(iii) $\underline{u}$ acts locally nilpotently on $\mathcal{P}$.

Proof. (i) already proved (lemma 2 and remark).

(ii) We have $\underline{u} = \underline{u}_0 + \underline{u}_1 + \cdots + \underline{u}_r$.

To verify that $\underline{u}$ is a Lie algebra, we only need check that $[\underline{u}_0, \underline{u}_1]$ vanishes at zero, since $[\underline{u}_i, \underline{u}_j] \subseteq \underline{n}_{i+j}$ and $\underline{u}_k = \underline{n}_k$ when $k \geq 2$. But $\underline{u}_0$ is spanned by vector fields $\xi^a \partial_i$, with ξ^a non-linear. The bracket with any vector field thus vanishes at zero.

To verify that $\underline{u}$ is an ideal, we first observe that the descriptions of $\underline{r}_0$ and $\underline{u}_0$ given in lemmas 1 and 3 make it clear that

$$[\underline{r}_0, \underline{u}_k] \subseteq \underline{u}_k \ , \qquad 1 \leq k \leq r \ .$$

By homogeneity, $[\underline{n}_\pm, \underline{u}_k] \subseteq \underline{n}_{k\pm 1}$. Since $\underline{u}_k = \underline{n}_k$ for $k \geq 2$, we only need check certain cases with $k = 0,1,2$.

Recall that $\underline{n}_+ = \mathcal{D}_1$. Since the vector fields in $\underline{u}_0$ have non-linear coefficients, we see that $[\underline{n}_+, \underline{u}_0]$ vanishes at 0, and hence

$$[\underline{n}_+, \underline{u}_0] \subseteq \underline{u}_1 \ .$$

The calculation made in the proof of Theorem A.2 shows that

$$[\underline{n}_-, \underline{u}_0] = 0 \ .$$

Furthermore, $\underline{n}_- = V_1^* \underline{h}$, so the vector fields in $\underline{n}_-$ have quadratic coefficients. It follows that $[\underline{n}_-, \underline{u}_2]$ vanishes at 0 , and that $[\underline{n}_-, \underline{u}_1]$ vanishes to second order at 0 . This shows, by the descriptions of $\underline{u}_0$ and $\underline{u}_1$ above, that

$$[\underline{n}_-, \underline{u}_2] \subseteq \underline{u}_1, \quad [\underline{n}_-, \underline{u}_1] \subseteq \underline{u}_0 \quad .$$

This completes the proof that $\underline{u}$ is an ideal

(iii) Let $P_m = \Sigma\, H_k \ (k \leq m)$ be the filtration on P associated with the decomposition $V = V_1 \oplus \cdots \oplus V_r$. We shall prove that

$$\underline{u}^m \,(P_m) = 0 \ .$$

Consider first the action of u_o . By lemma 3 we obtain

$$\underline{u}_o^{\ell} \,(H_m) \subseteq \Sigma\, H_{k_1} \cdots H_{k_{\ell+1}} \quad ,$$

where the sum is over all $k_1, \ldots, k_{\ell+1}$ such that $k_i < m$ and $\Sigma\, k_i = m$. Since $\underline{u}_o \,(H_1) = 0$, it follows that

$$\underline{u}_o^m \,(H_m) = 0 \quad .$$

Next, we have $\underline{u}_k \ P_m \subseteq P_{m-k}$, so that

$$\underline{u}_1^{k_1} \cdots \underline{u}_r^{k_r} \ P_m \subset P_{m-k} \quad ,$$

with $k = k_1 + \cdots + k_r$. Finally, the commutation relations $[\underline{u}_i, \underline{u}_j] \subseteq \underline{u}_{i+j}$ show that it suffices to consider ordered products

$$\underline{u}_o^{k_o} \, \underline{u}_1^{k_1} \cdots \underline{u}_r^{k_r} \ P_m \quad ,$$

with $k_o + k_1 + \cdots + k_r \geq m$. By the above, this is zero.

Since $V^* \subset P_r$, we obtain, in particular, a faithful representation of $\underline{u}$ on P_r by nilpotent transformations. Hence $\underline{u}$ is a nilpotent Lie algebra. This finishes the proof of (ii) and (iii) .

A.4 Birational Transformations Let $R = \mathrm{Rat}(V)$ be the field of rational functions on V, i.e. the quotient field of the integral domain P. We shall denote by $\mathrm{Aut}(R)$ the group of automorphisms of R (We require that $\varphi(\lambda) = \lambda$ for any scalar λ and automorphism φ.) There is a natural inclusion

$$\mathrm{Aut}(P) \subset \mathrm{Aut}(R) ,$$

where $\varphi(f/g) = \varphi f/\varphi g$, for $\varphi \in \mathrm{Aut}(P)$ and $f,g \in P$.

If $\xi_1,\ldots,\xi_n$ is a basis for V^*, then we can identify R with $\mathbb{C}(\xi_1,\ldots,\xi_n)$. A transformation $\varphi \in \mathrm{Aut}(R)$ is determined by its action on the generators $\xi_1,\ldots,\xi_n$, and can be expressed as

$$\varphi(\xi_i) = F_i(\xi_1,\ldots,\xi_n) ,$$

where F_i is a rational function in n variables. Conversely, given rational functions $F_1,\ldots,F_n$, then these equations define a unique homomorphism φ of R into R (we view R as an algebra over $\mathbb{C}$). This homomorphism will be an automorphism precisely when we can solve these equations rationally, i.e. find rational functions $G_1,\ldots,G_n$ such that

$$\xi_i = G_i(\varphi(\xi_1),\ldots,\varphi(\xi_n)) .$$

Hence the elements of $\mathrm{Aut}(R)$ are called birational transformations

Let $\underline{m}$ be the Lie algebra constructed in § 2, relative to a choice of a positive integral element $H \in \underline{h}$. We would like to "exponentiate" $\underline{m}$ in $\mathrm{Aut}(R)$. Now the group $\mathrm{Aut}(R)$ is not a Lie group, in any reasonable sense.

For example, as we noted in § A.1, a derivation of R (or P) does not necessarily generate a one-parameter group of automorphisms of R. Thus there is no functor mapping Lie subalgebras of derivations of R to Lie subgroups of $\mathrm{Aut}(R)$. For the algebra $\underline{m}$, however, we already know that the vector fields in $\underline{m}$ which are eigenvectors for ad $\underline{h}$ generate rational flows (in fact, those whose roots λ are non-negative generate polynomial flows). Hence it is reasonable to expect that every element of $\underline{m}$ generates a one-parameter group of automorphisms of R. We shall verify this by an explicit construction, using the information about the structure of $\underline{m}$ obtained in § A.3.

Recall that by Theorem A.3,

$$\underline{m} = \underline{n}_- \oplus \underline{r}_0 \oplus \underline{n}_+ \oplus \underline{u} \quad .$$

If $X \in \underline{u}$, then e^X is defined as an automorphism of the polynomial algebra. We set

$$U = \{e^X : X \in \underline{u}\} \quad .$$

By the proof of Theorem I.1.4, it follows (using the local nilpotency of $\underline{u}$ on P) that the map $X \to e^X$ is a bijection from $\underline{u}$ to U, and defines the structure of a complex, simply-connected nilpotent Lie group on U.

Let N_+ be the group of transformations $\{\rho_x : x \in V_1\}$, where

$$\rho_x f(v) = f(v+x) \;, \qquad f \in R \quad .$$

The map $\partial_x \mapsto \rho_x$ defines a bijection from $\underline{n}_+$ to N_+, and N_+ is isomorphic to the vector group V_1.

Let N_- be the group of transformations $\{\sigma_\xi : \xi \in V_1^*\}$, where σ_ξ is determined from its action on the linear functions by the formulas

$$\sigma_\xi(\zeta) = \zeta(1+\xi)^{-k} \;, \qquad \zeta \in V_k^*$$

The map $\xi H \mapsto \sigma_{-\xi}$ defines a bijection from $\underline{n}_-$ onto N_-, and N_- is isomorphic to the vector group V_1^*. (Note that the flow generated by the vector field ξH is $\sigma_{-t\xi}$, and σ_ξ is a birational transformation, by Remark A.1.5.)

To construct the subgroup R_0 corresponding to the subalgebra $\underline{r}_0$, our first impulse, based on lemma A.3.1, would be to take the direct product of the groups $GL(V_k)$, with the action on functions being induced by the linear action on V . This guess is indeed correct for $k \geq 2$. However, for $k = 1$ the action of $GL(V_1)$ must be "twisted" (cf. remark after lemma A.3.2).

Let G_k , $k \geq 2$, be the group of automorphisms $f \mapsto f \circ A^{-1}$, with $A \in GL(V_k)$ acting as the identity on the subspaces V_j , $j \neq k$. Thus $G_k \simeq GL(V_k)$. For the case $k = 1$, we let G_1 be the group of automorphisms whose action on <u>linear</u> functions is

$$(*) \qquad \zeta \mapsto (\det A)^{-k} \zeta \circ A^{-1} ,$$

when $\zeta \in V_k^*$. Here $A \in GL(V_1)$, and A is extended to act as the identity on the subspaces V_k , $k > 1$. Thus

$$G_1 \simeq GL(V_1)/D ,$$

where D is the finite central subgroup $\{\omega I : \omega^{q+1} = 1 \text{ and } \omega^{kq} = 1 \text{ for all } k \geq 2 \text{ such that } V_k \neq 0\}$. (Here $q = \dim V_1$) . It will be convenient to view $GL(V_1)$ and D as subgroups of $SL(q+1,\mathbb{C})$ via the embedding

$$A \mapsto \begin{pmatrix} A & 0 \\ 0 & (\det A)^{-1} \end{pmatrix} , \qquad A \in GL(V_1) .$$

Now define $R_0 = G_1 \times G_2 \times \cdots \times G_r$ (direct product as groups) , and let G be the subgroup of $\mathrm{Aut}(R)$ generated by N_- , R_0 , and N_+ . We shall show that G is isomorphic to Γ/D , where Γ is the direct product

$$\Gamma = SL(q+1,\mathbb{C}) \times GL(V_2) \times \cdots \times GL(V_r) , \quad (q = \dim V_1) .$$

For each factor $GL(V_k)$, $k \geq 2$, we already have an isomorphism with G_k . Furthermore, the groups $N_\pm$ and G_k , $k \geq 2$ mutually commute. Thus we only need to construct an isomorphism between $SL(q+1,\mathbb{C})/D$ and the group generated by N_- , G_1 , and N_+ .

We shall view $\Gamma_1 = SL(q+1,\mathbb{C})$ as the matrices in block form

$$(**) \qquad g = \begin{pmatrix} A & x \\ \xi & \lambda \end{pmatrix} , \qquad \det(g) = 1 ,$$

where $x \in V_1$, $\xi \in V_1^*$, $A \in End(V_1)$, and $\lambda \in \mathbb{C}$. The condition $\det(g) = 1$ can be expressed in terms of this decomposition as

$$\lambda \det A - < \xi, adj(A)x > = 1 ,$$

where adj A is the "adjugate" of A (transposed cofactor matrix). Write

$$\tilde{A} = adj(A) .$$

(Recall that $\tilde{A}A = A\tilde{A} = (\det A)\, I$.) We can imbed the vector groups V_1, V_1^* into $SL(q+1,\mathbb{C})$ as the subgroups

$$V_+ = \left\{ \begin{pmatrix} I & x \\ 0 & 1 \end{pmatrix} , \quad x \in V_1 \right\}$$

$$V_- = \left\{ \begin{pmatrix} I & 0 \\ \xi & 1 \end{pmatrix} , \quad \xi \in V_1^* \right\} .$$

Thus $V_\pm$ are commutative subgroups of Γ_1, which are normalized by $GL(V_1)$.

Theorem 1. There is a unique homomorphism π from Γ onto G whose restriction to $V_\pm$ is

$$\pi \begin{pmatrix} I & x \\ 0 & 1 \end{pmatrix} = \rho_x , \qquad \pi \begin{pmatrix} I & 0 \\ \xi & 1 \end{pmatrix} = \sigma_\xi ,$$

and whose restriction to $GL(V_k)$ is the homomorphism onto G_k defined above. The kernel of π is D.

Proof On the basis of the previous discussion, the only problem is to define π on Γ_1. Consider the subset of Γ_1 where $\det A \neq 0$ in $(**)$. On this set we have the unique factorisation

$$\begin{pmatrix} A & x \\ \xi & \lambda \end{pmatrix} = \begin{pmatrix} I & 0 \\ \xi A^{-1} & 1 \end{pmatrix} \begin{pmatrix} A & 0 \\ 0 & (\det A)^{-1} \end{pmatrix} \begin{pmatrix} I & A^{-1}x \\ 0 & 1 \end{pmatrix} .$$

Hence if π exists, it is given on this set by

$$(\#) \qquad \pi \begin{pmatrix} A & x \\ \xi & \lambda \end{pmatrix} = \sigma_{\xi A^{-1}} \; \pi(A) \; \rho_{A^{-1}x} .$$

Since the subgroups $V_\pm$ and $GL(V_1)$ generate Γ_1, formula (#) determines π uniquely, if such a homomorphism exists. Since Γ_1 is simply-connected (this is the only point at which we will need that the scalar field be $\mathbb{C}$), it suffices, for the existence of π, to show that (#) defines a local homomorphism from Γ_1 to $\mathrm{Aut}(R)$.

The adjoint action of $GL(V_1)$ on $V_\pm$ is $x \mapsto (\det A)Ax$, $\xi \mapsto (\det A)^{-1} \xi A^{-1}$, for $x \in V_1$, $\xi \in V_1^*$, as a matrix caculation shows. Since this agrees with the adjoint action of G_1 on $N_\pm$, this reduces the problem to the following:

Given $x \in V_1$, $\xi \in V_1^*$, we form the product

$$\begin{pmatrix} I & x \\ 0 & 1 \end{pmatrix} \begin{pmatrix} I & 0 \\ \xi & 1 \end{pmatrix} = \begin{pmatrix} A & x \\ \xi & 1 \end{pmatrix} ,$$

where $A = I + x \otimes \xi$ is the linear transformation sending $y \to y + \langle \xi, y \rangle x$. (If x, ξ are near 0, then A is invertible.). We must show that in this case

$$(?) \qquad \rho_x \sigma_\xi = \sigma_{\xi A^{-1}} \; \pi(A) \; \rho_{A^{-1}x} .$$

Both sides of (?) are automorphisms of R, so it suffices to calculate their action on a linear function $\zeta \in V_k^*$. We find that

$$\rho_x \sigma_\xi (\zeta) = (\zeta + \langle \zeta, x \rangle)(1 + \langle \xi, x \rangle + \xi)^{-k} .$$

On the other hand, we have

$$\sigma_{\xi A^{-1}} \pi(A)\rho_{A^{-1}x}(\zeta) = \frac{\zeta A^{-1} + < \zeta,\tilde{A}x > \xi A^{-1} + < \zeta,\tilde{A}x >}{(\xi\tilde{A} + \det A)^k} .$$

But $\det(I + x \otimes \xi) = 1 + < \xi,x >$, and if $< \xi,x > \neq -1$, we can write

$$A^{-1} = I - (1 + < \xi,x >)^{-1} x \otimes \xi$$

$$\tilde{A} = (\det A) A^{-1} = (1 + < \xi,x >) I - x \otimes \xi$$

Using these formulas, we easily verify (?) , and hence establish the existence of π .

Obviously $\operatorname{Ker} \pi \subset \Gamma_1$, and hence $\operatorname{Ker} \pi \subset \{\omega I : \omega^{q+1} = 1\}$, since it is a normal subgroup. This implies that $\operatorname{Ker}(\pi) = D$, Q.E.D.

Remark. If $g = \begin{pmatrix} A & x \\ \xi & \lambda \end{pmatrix} \in \Gamma_1$,

then the calculation made in the proof of theorem 1 shows that

$$(\#\#) \qquad \pi(g)\zeta = \frac{\zeta A^{-1} + < \zeta,\tilde{A}x > \xi A^{-1} + < \zeta,\tilde{A}x >}{(\xi\tilde{A} + \det A)^k} ,$$

if $\zeta \in V_k^*$. (Here A is extended to be I on V_k^* for $k > 1$, by definition.)

This formula remains meaningful even when A is singular. To see this, recall that if $\det A = 0$, then the condition $\det g = 1$ forces

$$< \xi,\tilde{A}x > = -1$$

In particular, $\xi\tilde{A} \neq 0$, so the denominator of (##) is never identically zero.

For the numerator, we write

$$\zeta A^{-1} + < \zeta,\tilde{A}x > \xi A^{-1} = (\det A)^{-1}(\zeta\tilde{A} + < \zeta,\tilde{A}x > \xi\tilde{A})$$

If $\zeta \in V_k^*$ with $k > 1$, then $< \zeta,\tilde{A}x > = 0$ and $\zeta A^{-1} = \zeta$, by definition. To examine the behavior on V_1^* , pick a basis $\{x_i\}$ for V_1 and dual basis

$\{\xi_i\}$ for V_1^*. Using the condition $\det g = 1$, we calculate that when $\zeta = \xi_i$,

$$\xi_i\tilde{A} + \langle \xi_i, \tilde{A}x \rangle \xi\tilde{A} = \lambda(\det A)\, \xi_i\tilde{A}$$

$$+ \Sigma\, \varphi_\ell(g)\xi_\ell \quad ,$$

where

$$\varphi_\ell(g) = \sum_{j,k} \langle \xi_j, x \rangle \langle \xi, x_k \rangle \det \begin{bmatrix} c_{ji} & c_{\ell i} \\ c_{jk} & c_{\ell k} \end{bmatrix} .$$

Here $[c_{ij}]$ is the cofactor matrix of A relative to this basis. But it is a classical fact in determinant theory ("Jacobi's formulas") that the "compound determinants" appearing in the formula for φ_ℓ are divisible by $\det A$. (For a non-computational proof, use the fact that $\det A = 0 \Rightarrow \operatorname{rank} \tilde{A} \leq 1 \Rightarrow$ all 2×2 minors of $\tilde{A}$ vanish $\Rightarrow \varphi_\ell(g) = 0$. Since det is an irreducible polynomial, it must thus divide φ_ℓ.)

This shows that the numerator in (##) is everywhere regular on Γ_1. By analytic continuation, the extension of π from $V_-G_1V_+$ to Γ is in fact given by (##).

To complete our construction of a group of automorphisms corresponding to the Lie algebra $\underline{m}$, we only need to put together the groups G and U, as follows:

Theorem 2 G normalises U, and $G \cap U = \{1\}$. Hence $M = G\,U$ is a subgroup of $\operatorname{Aut}(\mathcal{R})$.

Proof From the proof of Theorem A.3, we see that $\operatorname{ad}(\underline{n}_\pm)$ acts nilpotently on $\underline{u}$, and we verify easily that $\operatorname{Ad}(N_\pm)$ stabilises $\underline{u}$. By the structure of $\underline{u}$ determined in Lemma A.3.3 we find that $\operatorname{Ad}(R_0)$ also stabilises $\underline{u}$. The passage from $\underline{u}$ to U follows directly from these calculations and the local nilpotence of $\underline{u}$ on $\mathcal{P}$

To prove that $G \cap U = \{1\}$, we observe first that since $G \cap U \subset \operatorname{Aut}(\mathcal{P})$, one has

$$G \cap U \subseteq R_0N_+U \ .$$

(In formula (##) above, if $\xi \neq 0$ then $\pi(g)\zeta$ is not a polynomial function on V.)

Set $T = N_+U$. From lemmas 2 and 3 of the previous section we see that $\underline{n}_+ + \underline{u}$ is a nilpotent Lie algebra which acts locally nilpotently on P. Hence

$$T = \exp(\underline{n}_+ + \underline{u}) \quad .$$

On the other hand, we evidently have

$$\underline{n}_+ + \underline{u} = \underline{u}_0 + \underline{n}_1 + \underline{n}_2 + \ldots + \underline{n}_r \quad .$$

If we define the spaces H_m^o of non-linear homogeneous polynomials of degree m as in lemma 3, and set

$$Q_m = H_m^o + P_{m-1} \quad , \quad m \geq 1 \quad ,$$

then we can describe the Lie algebra $\underline{n}_+ + \underline{u}$ as

$$\underline{n}_+ + \underline{u} = \{X \in \mathrm{Der}(P) : X P_m \subseteq Q_m \text{ for all } m\} \ .$$

(cf. lemma A.3.3 and § I.1.3). Exponentiating this description, we thus can characterise the group T as

$$T = \{\varphi \in \mathrm{Aut}(P) : (\varphi - 1) P_m \subseteq Q_m \text{ for all } m\} \ .$$

Now the group R_o acts linearly on V^*, leaving each subspace V_k^* invariant. The above description of T makes it evident that

$$R_o \cap T = \{1\} \ .$$

Finally, we have $N_+ \cap U = \{1\}$ because the exponential map is a bijection from $\underline{n}_+ + \underline{v}$ onto T. This completes the proof.

<u>Example</u> Let us return to the examples (dim V = 2) at the end of § A.2, and employ the same notation. The algebra $\underline{m}$ is determined by a choice of positive integer n. When n = 1, then M = SL(3), acting projectively on {x,y}. (View x,y as inhomogeneous coordinates for $\mathbb{P}^2$.)

When $n > 1$, then $M = G\,U$, where

$$G \simeq SL(2) \times GL(1)$$

Here $g = \left\{\begin{pmatrix} a & b \\ c & d \end{pmatrix},\ c_0\right\} \in G$ acts by the birational transformation

$$\begin{cases} x \to (b + dx)(a + cx)^{-1} \\ y \to c_0 y\,(a + cx)^{-n} \end{cases}$$

The group U consists of all transformations

$$\begin{cases} x \to x \\ y \to y + c_1\,x + \ldots + c_n\,x^n \end{cases}$$

$(c_i \in \mathbb{C})$. The group M is the classical "Jonquières group of order n."

Comments and references for Appendix

The study of finite-dimensional Lie subgroups of the (infinite-dimensional) group of birational transformations of an affine space has a long history; cf. Fano [1]. The classical "Jonquières groups" in two variables occurred in the classification by Enriques of all finite-dimensional groups of birational transformations in two variables. They were studied in more detail by Mohrmann [1] and Godeaux [1], and "automorphic functions" on these groups were considered by Myrberg [1]; cf. the survey article by Coble [1].

In recent years the subject has been greatly extended by Demazure [1] and Vinberg [1]. The algebras and groups we construct here furnish a class of examples for Demazure's general theory of "Enriques systems". Several of our proofs are special cases of his general methods. Since there exists no classification of Enriques systems, as contrasted to the classification of root systems for semi-simple algebras, it is perhaps useful to have such examples constructed explicitly. The fact that the Lie algebras $\underline{m}$ are maximal seems to be new. The subalgebra $\underline{m}_o$ of vector fields homogeneous of degree zero has appeared also in Arnol'd [1], in connection with the classification of normal forms for smooth functions at a critical point. For "Jacobi's formulas", used in the Remark in § 4, cf. Hodge-Pedoe [1], Chap. 2, § 8.

Bibliography

L. Auslander, J. Brezin, and R. Sacksteder

[1] A method in metric diophantine approximation, J. Differential Geometry 6(1972), 479-496.

V.I. Arnol'd

[1] Critical points of smooth functions and their normal forms, Uspekhi Math. Nauk 30(1975), 3-66.

F.A. Berezin

[1] Some remarks on the associative envelopes of Lie algebras, Funkt. Analiz i ego Pril. 1(1967), 1-14.

L. Bers, F. John, and M. Schechter

[1] Partial differential equations, in Lectures in Applied Mathematics Vol. III, Interscience, New York, 1964.

G. Birkhoff

[1] Representability of Lie algebras and Lie groups by matrices, Annals of Math. 38(1937), 526-532.

N. Bourbaki

[1] Elêments de mathêmatique XXXIV, Groupes et algêbres de Lie, Hermann, Paris, 1968.

T. Bloom and I. Graham

[1] A geometric characterisation of points of type m on real submanifolds of $\mathbb{C}^n$, J. Differential Geometry (to appear).

P. Cartier

[1] Quantum mechanical commutation relations and theta functions, in Algebraic Groups and discontinuous subgroups, Amer. Math. Soc., 1966, 361-383.

[2] Vecteurs diffêrentiables dans les reprêsentations unitaires des groupes de Lie, Sêminaire Bourbaki 27(1974/75), N$^{\underline{o}}$ 454.

S.S. Chern and J. Moser

[1] Real hypersurfaces in complex manifolds,Acta Math. 133(1974), 219-272.

A.B. Coble

[1] Cremona transformations and applications to algebra, geometry, and modular functions, Bull. Amer. Math. Soc. 28(1922), 329-364.

R. Coiffman and G. Weiss

[1] Analyse Harmonique Non-Commutative sur Certains Espaces Homogenes, Lecture Notes in Mathematics 242, Springer-Verlag, 1971.

N. Conze

[1] Algèbres d'opérateurs différentials et quotients des algèbres enveloppantes, Bull. Soc. Math. France 102(1974), 379-415.

M. Cotlar and C. Sadosky

[1] On Quasi-homogeneous Bessel Potential Operators, in Singular Integrals, Amer. Math. Soc., 1966, 275-287.

M. Demazure

[1] Sous-groupes algébriques de rang maximum du group de Cremona, Ann. Scient. Ec. Norm. Sup.(4), t.3(1970), 507-588. (cf. Sem. Bourbaki, 1971/72, Exposé 413)

J. Dixmier

[1] Cohomologie des algèbres de Lie nilpotentes, Acta Scientiarum Mathematicarum 16(1955), 246-250.

[2] Algèbres enveloppantes, Paris, Gauthier-Villars, 1974 (Cahiers scientifiques 37).

J. Dyer

[1] A nilpotent Lie algebra with nilpotent automorphism group, Bull. Amer. Math. Soc. 76(1970), 52-56.

G. Fano

[1] Kontinuierliche Geometrische Gruppen, Encyclopaedie der Math. Wiss. III_1, Geometrie, 4b, § 21-22.

G. Favre

[1] Systèmes de poids sur une algèbre nilpotente, Manuscripta Math. 9(1973), 53-90.

G.B. Folland

[1] A fundamental solution for a subelliptic operator, Bull. Amer. Math. Soc. 79(1973), 373-376.

[2] Subelliptic estimates and function spaces on nilpotent Lie groups, Arkiv för matematik 13(1975), 161-208.

G.B. Folland and J.J. Kohn

[1] The Neumann Problem for the Cauchy-Riemann complex, Ann. of Math. Studies #75, Princeton Univ. Press, Princeton, N.J., 1972.

G.B. Folland and E.M. Stein

[1] Estimates for the $\bar{\partial}_b$ complex and analysis on the Heisenberg group, Comm. Pure Appl. Math. 27(1974), 429-522.

B. Gaveau

[1] Solutions fondamentales, représentations, et estimées sous-elliptiques pour les groupes nilpotent d'ordre 2, C.R. Acad. Sc. Paris 282(1976), A 563-566.

S.G. Gindikin

[1] Analysis in homogeneous domains, Russian Math. Surveys 19(1964), 1-90.

L. Godeaux

[1] Les transformations birationnelles du plan Mem, des Sc. Math. XXII, Gauthier-Villars, Paris, 1927.

[2] Sur les transformations birationelles de Jonquières de l'éspace, Acad. roy. Belgique, 1922.

Goodman, R.

[1] One-parameter groups generated by operators in an enveloping algebra, J. Functional Analysis 6(1970), 218-236.

[2] Complex Fourier analysis on nilpotent Lie groups, Trans. Amer. Math. Soc. 160(1971), 373-391.

[3] Differential operators of infinite order on a Lie group II, Indiana Univ. Math. J. 21(1971), 383-409.

[4] Some regularity theorems for operators in an enveloping algebra, J. Differential Equations 10(1971), 448-470.

[5] On the boundedness and unboundedness of certain convolution operators on nilpotent Lie groups, Proc. Amer. Math. Soc. 39(1973), 409-413.

[6] Filtrations and asymptotic automorphisms on nilpotent Lie groups, J. Differential Geometry (to appear).

[7] Lifting vector fields to nilpotent Lie groups, (to appear).

V.V. Grušin

[1] On a class of hypoelliptic operators, Mat. Sbornik 83 (125) (1970), 456-473 (=Math. USSR Sbornik 12(1970), 458-476).

[2] Hypoelliptic differential equations and pseudodifferential operators with operator-valued symbols, Mat. Sbornik 88 (130) (1972), 504-521 (=Math. USSR Sbornik 17(1972), 497-514).

Y. Guivarc'h

[1] Croissance polynomiale et pēriodes des fonctions harmoniques, Bull. Soc. Math. France 101(1973), 333-379.

S. Helgason

[1] Differential Geometry and Symmetric Spaces, Academic Press, New York, 1962.

[2] A duality for symmetric spaces with applications to group representations, Advances in Math. 5(1970), 1-154.

G. Hochschild

[1] The Structure of Lie Groups, Holden-Day, San Francisco, 1965.

W.V.D. Hodge and D. Pedoe

[1] Methods of Algebraic Geometry, I, Cambridge University Press, 1968.

L. Hörmander

[1] Hypoelliptic second-order differential equations, Acta Math. 119(1968), 147-171.

A. Hulanicki

[1] Commutative subalgebra of $L^1(G)$ associated with a subelliptic operator on a Lie group, Bull. Amer. Math. Soc. 81(1975), 121-124.

N. Jacobson

[1] Lie Algebras, Interscience, New York, 1962.

J. Jenkins

[1] Dilations, gauges, and growth in nilpotent Lie groups, preprint, June, 1975.

K. Johnson and N. Wallach

[1] Composition series and intertwining operators for the spherical principal series, Bull. Amer. Math. Soc. 78(1972), 1053-1059.

R.W. Johnson

[1] Homogeneous Lie algebras and expanding automorphisms, preprint, 1974.

P. Jørgensen

[1] Representations of differential operators on a Lie group, J. Functional Analysis 20(1975), 105-135.

A.W. Knapp

[1] A Szegö kernel for discrete series, Proc. Inter. Cong. Math. Vancouver (1974), Vol. 2, 99-104.

A.W. Knapp and E.M. Stein

[1] Intertwining operators for semisimple groups, Annals of Math. (2) 93 (1971), 489-578.

[2] Singular integrals and the principal series III, Proc. Nat. Acad. Sc. USA 71(1974), 4622-4624

[3] Singular integrals and the principal series IV, Proc. Nat. Acad. Sci. USA 72(1975), 2459-2461.

J.J. Kohn

[1] Pseudo-differential operators and hypoellipticity, in Partial Differential Equations, Amer. Math. Soc. (1973), 61-70.

[2] Integration of complex vector fields, Bull. Amer. Math. Soc. 78(1972), 1-11.

[3] Boundary behaviour of $\bar{\partial}$ on weakly pseudo-convex manifolds of dimension two, J. Differential Geometry 6(1972), 523-542.

A. Korânyi

[1] The Poisson integral for generalized half-planes and bounded symmetric domains, Annals of Math (2) 82(1965), 332-350.

A. Korânyi and S. Vâgi

[1] Singular integrals in homogeneous spaces and some problems in classical analysis, Ann. Scuola Norm. Sup. Pisa, Classe di Scienze 25(1971), 575-648.

A. Korânyi, S. Vâgi, and G. Welland

[1] Remarks on the Cauchy integral and the conjugate function in generalized half-planes, J. Math. and Mech. 19(1970), 1069-1081.

R. Kunze and E.M. Stein

[1] Uniformly bounded representations III, Amer. J. of Math. 89(1967), 385-442.

H. Mohrmann

[1] Über die automorphe Kollineationsgruppe des rationalen Normalkegels n-ter Ordnung, Rendic. del Circolo Mat. di Palermo 31(1911), 170-200.

C.C. Moore

[1] Representations of solvable and nilpotent groups and harmonic analysis on nil and solvmanifolds, in Harmonic Analysis on Homogeneous Spaces, Amer. Math. Soc. (1972), 3-44.

P. Müller-Römer

[1] Kontrahierende Erweiterungen und kontrahierbare Gruppen, J.f.d. Reine und Angewandte Mathematik 283/284 (1976) 238-264.

P.J. Myrber

[1] Über die automorphen Funktionen bei einer Klasse Jonquièresscher Gruppen zweier Veränderlichen, Math. Zeitschrift 21(1924), 224-253.

A. Nijenhuis and R.W. Richardson, Jr.

[1] Deformations of algebraic structures, Bull. Amer. Math. Soc. 70(1964), 406-411.

R.D. Ogden and S. Vãgi

[1] Harmonic analysis and H^2 functions on Siegel domains of type II, Proc. Nat. Acad. Sci. USA 69(1972), 11-14.

N.S. Poulsen

[1] C^∞ vectors and intertwining bilinear forms for representations of Lie groups, J. Functional Analysis 9(1972), 87-120.

G. Rauch

[1] Variation d'algèbres de Lie résolubles, C.R. Acad. Sci. Paris 269(1969), A 685-687.

[2] Variations d'algèbres de Lie, Pub. Math. Univ. Paris XI (Orsay), N$^{\underline{o}}$ 32, 1973.

B. Reed

[1] Representations of solvable Lie algebras, Michigan Math. J. 16(1969), 227-233.

C. Rockland

[1] Hypoellipticity on the Heisenberg group-representation theoretic criteria (preprint).

L.P. Rothschild and E.M. Stein

[1] Hypoelliptic differential operators and nilpotent groups, Acta Math. (to appear).

G. Schiffmann

[1] Intégrales d'entrelacement et fonctions de Whittaker, Bull. Soc. Math. France 99(1971), 3-72.

E.M. Stein

[1] "Some problems in harmonic analysis suggested by symmetric spaces and semi-simple Lie groups," Actes du Congrès International des Mathématiciens, Nice, 1970, Tom. 1, pp. 173-189.

[2] Boundary behavior of holomorphic functions of several complex variables, Mathematical Notes, Princeton Univ. Press, 1972.

R.S. Strichartz

[1] Singular integrals on nilpotent Lie groups, Proc. Amer. Math. Soc. 53(1975), 367-374.

N. Tanaka

[1] On differential systems, graded Lie algebras and pseudo-groups, J. Math. Kyoto Univ. 10(1970), 1-82.

[2] A differential-geometric study on strongly pseudo-convex manifolds, Lectures in Math. # 9, Dept. of Math., Kyoto Univ., 1975.

F. Trèves

[1] <u>Topological Vector Spaces, Distributions, and Kernels</u>, New York, Academic Press, 1967.

M. Vergne

[1] Cohomologie des algèbres de Lie nilpotentes; application à l'étude de la variété des algèbres de Lie nilpotentes, Bull. Soc. Math. France 98(1970), 81-116.

E.B. Vinberg

[1] Algebraic groups of transformations of maximal rank, Math. Sbornik 88 (130) (1972), 493-503. (=Math. USSR Sbornik 17(1972), 487-496).

N.R. Wallach

[1] <u>Harmonic Analysis on Homogeneous Spaces</u>, Marcel Dekker, New York, 1973.

G. Warner

[1] <u>Harmonic Analysis of Semi-Simple Lie Groups</u>, Springer-Verlag, Berlin, 1972.

Subject Index

Vol. 399: Functional Analysis and its Applications. Proceedings 1973. Edited by H. G. Garnir, K. R. Unni and J. H. Williamson. II, 584 pages. 1974.

Vol. 400: A Crash Course on Kleinian Groups. Proceedings 1974. Edited by L. Bers and I. Kra. VII, 130 pages. 1974.

Vol. 401: M. F. Atiyah, Elliptic Operators and Compact Groups. V, 93 pages. 1974.

Vol. 402: M. Waldschmidt, Nombres Transcendants. VIII, 277 pages. 1974.

Vol. 403: Combinatorial Mathematics. Proceedings 1972. Edited by D. A. Holton. VIII, 148 pages. 1974.

Vol. 404: Théorie du Potentiel et Analyse Harmonique. Edité par J. Faraut. V, 245 pages. 1974.

Vol. 405: K. J. Devlin and H. Johnsbråten, The Souslin Problem. VIII, 132 pages. 1974.

Vol. 406: Graphs and Combinatorics. Proceedings 1973. Edited by R. A. Bari and F. Harary. VIII, 355 pages. 1974.

Vol. 407: P. Berthelot, Cohomologie Cristalline des Schémas de Caracteristique $p > o$. II, 604 pages. 1974.

Vol. 408: J. Wermer, Potential Theory. VIII, 146 pages. 1974.

Vol. 409: Fonctions de Plusieurs Variables Complexes, Séminaire François Norguet 1970–1973. XIII, 612 pages. 1974.

Vol. 410: Séminaire Pierre Lelong (Analyse) Année 1972–1973. VI, 181 pages. 1974.

Vol. 411: Hypergraph Seminar. Ohio State University, 1972. Edited by C. Berge and D. Ray-Chaudhuri. IX, 287 pages. 1974.

Vol. 412: Classification of Algebraic Varieties and Compact Complex Manifolds. Proceedings 1974. Edited by H. Popp. V, 333 pages. 1974.

Vol. 413: M. Bruneau, Variation Totale d'une Fonction. XIV, 332 pages. 1974.

Vol. 414: T. Kambayashi, M. Miyanishi and M. Takeuchi, Unipotent Algebraic Groups. VI, 165 pages. 1974.

Vol. 415: Ordinary and Partial Differential Equations. Proceedings 1974. XVII, 447 pages. 1974.

Vol. 416: M. E. Taylor, Pseudo Differential Operators. IV, 155 pages. 1974.

Vol. 417: H. H. Keller, Differential Calculus in Locally Convex Spaces. XVI, 131 pages. 1974.

Vol. 418: Localization in Group Theory and Homotopy Theory and Related Topics. Battelle Seattle 1974 Seminar. Edited by P. J. Hilton. VI, 172 pages 1974.

Vol. 419: Topics in Analysis. Proceedings 1970. Edited by O. E. Lehto, I. S. Louhivaara, and R. H. Nevanlinna. XIII, 392 pages. 1974.

Vol. 420: Category Seminar. Proceedings 1972/73. Edited by G. M. Kelly. VI, 375 pages. 1974.

Vol. 421: V. Poénaru, Groupes Discrets. VI, 216 pages. 1974.

Vol. 422: J.-M. Lemaire, Algèbres Connexes et Homologie des Espaces de Lacets. XIV, 133 pages. 1974.

Vol. 423: S. S. Abhyankar and A. M. Sathaye, Geometric Theory of Algebraic Space Curves. XIV, 302 pages. 1974.

Vol. 424: L. Weiss and J. Wolfowitz, Maximum Probability Estimators and Related Topics. V, 106 pages. 1974.

Vol. 425: P. R. Chernoff and J. E. Marsden, Properties of Infinite Dimensional Hamiltonian Systems. IV, 160 pages. 1974.

Vol. 426: M. L. Silverstein, Symmetric Markov Processes. X, 287 pages. 1974.

Vol. 427: H. Omori, Infinite Dimensional Lie Transformation Groups. XII, 149 pages. 1974.

Vol. 428: Algebraic and Geometrical Methods in Topology, Proceedings 1973. Edited by L. F. McAuley. XI, 280 pages. 1974.

Vol. 429: L. Cohn, Analytic Theory of the Harish-Chandra C-Function. III, 154 pages. 1974.

Vol. 430: Constructive and Computational Methods for Differential and Integral Equations. Proceedings 1974. Edited by D. L. Colton and R. P. Gilbert. VII, 476 pages. 1974.

Vol. 431: Séminaire Bourbaki – vol. 1973/74. Exposés 436–452. IV, 347 pages. 1975.

Vol. 432: R. P. Pflug, Holomorphiegebiete, pseudokonvexe Gebiete und das Levi-Problem. VI, 210 Seiten. 1975.

Vol. 433: W. G. Faris, Self-Adjoint Operators. VII, 115 pages. 1975.

Vol. 434: P. Brenner, V. Thomée, and L. B. Wahlbin, Besov Spaces and Applications to Difference Methods for Initial Value Problems. II, 154 pages. 1975.

Vol. 435: C. F. Dunkl and D. E. Ramirez, Representations of Commutative Semitopological Semigroups. VI, 181 pages. 1975.

Vol. 436: L. Auslander and R. Tolimieri, Abelian Harmonic Analysis, Theta Functions and Function Algebras on a Nilmanifold. V, 99 pages. 1975.

Vol. 437: D. W. Masser, Elliptic Functions and Transcendence. XIV, 143 pages. 1975.

Vol. 438: Geometric Topology. Proceedings 1974. Edited by L. C. Glaser and T. B. Rushing. X, 459 pages. 1975.

Vol. 439: K. Ueno, Classification Theory of Algebraic Varieties and Compact Complex Spaces. XIX, 278 pages. 1975

Vol. 440: R. K. Getoor, Markov Processes: Ray Processes and Right Processes. V, 118 pages. 1975.

Vol. 441: N. Jacobson, PI-Algebras. An Introduction. V, 115 pages. 1975.

Vol. 442: C. H. Wilcox, Scattering Theory for the d'Alembert Equation in Exterior Domains. III, 184 pages. 1975.

Vol. 443: M. Lazard, Commutative Formal Groups. II, 236 pages. 1975.

Vol. 444: F. van Oystaeyen, Prime Spectra in Non-Commutative Algebra. V, 128 pages. 1975.

Vol. 445: Model Theory and Topoi. Edited by F. W. Lawvere, C. Maurer, and G. C. Wraith. III, 354 pages. 1975.

Vol. 446: Partial Differential Equations and Related Topics. Proceedings 1974. Edited by J. A. Goldstein. IV, 389 pages. 1975.

Vol. 447: S. Toledo, Tableau Systems for First Order Number Theory and Certain Higher Order Theories. III, 339 pages. 1975.

Vol. 448: Spectral Theory and Differential Equations. Proceedings 1974. Edited by W. N. Everitt. XII, 321 pages. 1975.

Vol. 449: Hyperfunctions and Theoretical Physics. Proceedings 1973. Edited by F. Pham. IV, 218 pages. 1975.

Vol. 450: Algebra and Logic. Proceedings 1974. Edited by J. N. Crossley. VIII, 307 pages. 1975.

Vol. 451: Probabilistic Methods in Differential Equations. Proceedings 1974. Edited by M. A. Pinsky. VII, 162 pages. 1975.

Vol. 452: Combinatorial Mathematics III. Proceedings 1974. Edited by Anne Penfold Street and W. D. Wallis. IX, 233 pages. 1975.

Vol. 453: Logic Colloquium. Symposium on Logic Held at Boston, 1972–73. Edited by R. Parikh. IV, 251 pages. 1975.

Vol. 454: J. Hirschfeld and W. H. Wheeler, Forcing, Arithmetic, Division Rings. VII, 266 pages. 1975.

Vol. 455: H. Kraft, Kommutative algebraische Gruppen und Ringe. III, 163 Seiten. 1975.

Vol. 456: R. M. Fossum, P. A. Griffith, and I. Reiten, Trivial Extensions of Abelian Categories. Homological Algebra of Trivial Extensions of Abelian Categories with Applications to Ring Theory. XI, 122 pages. 1975.

Vol. 457: Fractional Calculus and Its Applications. Proceedings 1974. Edited by B. Ross. VI, 381 pages. 1975.

Vol. 458: P. Walters, Ergodic Theory – Introductory Lectures. VI, 198 pages. 1975.

Vol. 459: Fourier Integral Operators and Partial Differential Equations. Proceedings 1974. Edited by J. Chazarain. VI, 372 pages. 1975.

Vol. 460: O. Loos, Jordan Pairs. XVI, 218 pages. 1975.

Vol. 461: Computational Mechanics. Proceedings 1974. Edited by J. T. Oden. VII, 328 pages. 1975.

Vol. 462: P. Gérardin, Construction de Séries Discrètes p-adiques. »Sur les séries discrètes non ramifiées des groupes réductifs déployés p-adiques«. III, 180 pages. 1975.

Vol. 463: H.-H. Kuo, Gaussian Measures in Banach Spaces. VI, 224 pages. 1975.

Vol. 464: C. Rockland, Hypoellipticity and Eigenvalue Asymptotics. III, 171 pages. 1975.

Vol. 465: Séminaire de Probabilités IX. Proceedings 1973/74. Edité par P. A. Meyer. IV, 589 pages. 1975.

Vol. 466: Non-Commutative Harmonic Analysis. Proceedings 1974. Edited by J. Carmona, J. Dixmier and M. Vergne. VI, 231 pages. 1975.

Vol. 467: M. R. Essén, The Cos $\pi\lambda$ Theorem. With a paper by Christer Borell. VII, 112 pages. 1975.

Vol. 468: Dynamical Systems – Warwick 1974. Proceedings 1973/74. Edited by A. Manning. X, 405 pages. 1975.

Vol. 469: E. Binz, Continuous Convergence on C(X). IX, 140 pages. 1975.

Vol. 470: R. Bowen, Equilibrium States and the Ergodic Theory of Anosov Diffeomorphisms. III, 108 pages. 1975.

Vol. 471: R. S. Hamilton, Harmonic Maps of Manifolds with Boundary. III, 168 pages. 1975.

Vol. 472: Probability-Winter School. Proceedings 1975. Edited by Z. Ciesielski, K. Urbanik, and W. A. Woyczyński. VI, 283 pages. 1975.

Vol. 473: D. Burghelea, R. Lashof, and M. Rothenberg, Groups of Automorphisms of Manifolds. (with an appendix by E. Pedersen) VII, 156 pages. 1975.

Vol. 474: Séminaire Pierre Lelong (Analyse) Année 1973/74. Edité par P. Lelong. VI, 182 pages. 1975.

Vol. 475: Répartition Modulo 1. Actes du Colloque de Marseille-Luminy, 4 au 7 Juin 1974. Edité par G. Rauzy. V, 258 pages. 1975. 1975.

Vol. 476: Modular Functions of One Variable IV. Proceedings 1972. Edited by B. J. Birch and W. Kuyk. V, 151 pages. 1975.

Vol. 477: Optimization and Optimal Control. Proceedings 1974. Edited by R. Bulirsch, W. Oettli, and J. Stoer. VII, 294 pages. 1975.

Vol. 478: G. Schober, Univalent Functions – Selected Topics. V, 200 pages. 1975.

Vol. 479: S. D. Fisher and J. W. Jerome, Minimum Norm Extremals in Function Spaces. With Applications to Classical and Modern Analysis. VIII, 209 pages. 1975.

Vol. 480: X. M. Fernique, J. P. Conze et J. Gani, Ecole d'Eté de Probabilités de Saint-Flour IV-1974. Edité par P.-L. Hennequin. XI, 293 pages. 1975.

Vol. 481: M. de Guzmán, Differentiation of Integrals in R^n. XII, 226 pages. 1975.

Vol. 482: Fonctions de Plusieurs Variables Complexes II. Séminaire François Norguet 1974–1975. IX, 367 pages. 1975.

Vol. 483: R. D. M. Accola, Riemann Surfaces, Theta Functions, and Abelian Automorphisms Groups. III, 105 pages. 1975.

Vol. 484: Differential Topology and Geometry. Proceedings 1974. Edited by G. P. Joubert, R. P. Moussu, and R. H. Roussarie. IX, 287 pages. 1975.

Vol. 485: J. Diestel, Geometry of Banach Spaces – Selected Topics. XI, 282 pages. 1975.

Vol. 486: S. Stratila and D. Voiculescu, Representations of AF-Algebras and of the Group U (∞). IX, 169 pages. 1975.

Vol. 487: H. M. Reimann und T. Rychener, Funktionen beschränkter mittlerer Oszillation. VI, 141 Seiten. 1975.

Vol. 488: Representations of Algebras, Ottawa 1974. Proceedings 1974. Edited by V. Dlab and P. Gabriel. XII, 378 pages. 1975.

Vol. 489: J. Bair and R. Fourneau, Etude Géométrique des Espaces Vectoriels. Une Introduction. VII, 185 pages. 1975.

Vol. 490: The Geometry of Metric and Linear Spaces. Proceedings 1974. Edited by L. M. Kelly. X, 244 pages. 1975.

Vol. 491: K. A. Broughan, Invariants for Real-Generated Uniform Topological and Algebraic Categories. X, 197 pages. 1975.

Vol. 492: Infinitary Logic: In Memoriam Carol Karp. Edited by D. W. Kueker. VI, 206 pages. 1975.

Vol. 493: F. W. Kamber and P. Tondeur, Foliated Bundles and Characteristic Classes. XIII, 208 pages. 1975.

Vol. 494: A Cornea and G. Licea. Order and Potential Resolvent Families of Kernels. IV, 154 pages. 1975.

Vol. 495: A. Kerber, Representations of Permutation Groups II. V, 175 pages. 1975.

Vol. 496: L. H. Hodgkin and V. P. Snaith, Topics in K-Theory. Two Independent Contributions. III, 294 pages. 1975.

Vol. 497: Analyse Harmonique sur les Groupes de Lie. Proceedings 1973–75. Edité par P. Eymard et al. VI, 710 pages. 1975.

Vol. 498: Model Theory and Algebra. A Memorial Tribute to Abraham Robinson. Edited by D. H. Saracino and V. B. Weispfenning. X, 463 pages. 1975.

Vol. 499: Logic Conference, Kiel 1974. Proceedings. Edited by G. H. Müller, A. Oberschelp, and K. Potthoff. V, 651 pages 1975.

Vol. 500: Proof Theory Symposion, Kiel 1974. Proceedings. Edited by J. Diller and G. H. Müller. VIII, 383 pages. 1975.

Vol. 501: Spline Functions, Karlsruhe 1975. Proceedings. Edited by K. Böhmer, G. Meinardus, and W. Schempp. VI, 421 pages. 1976.

Vol. 502: János Galambos, Representations of Real Numbers by Infinite Series. VI, 146 pages. 1976.

Vol. 503: Applications of Methods of Functional Analysis to Problems in Mechanics. Proceedings 1975. Edited by P. Germain and B. Nayroles. XIX, 531 pages. 1976.

Vol. 504: S. Lang and H. F. Trotter, Frobenius Distributions in GL_2-Extensions. III, 274 pages. 1976.

Vol. 505: Advances in Complex Function Theory. Proceedings 1973/74. Edited by W. E. Kirwan and L. Zalcman. VIII, 203 pages. 1976.

Vol. 506: Numerical Analysis, Dundee 1975. Proceedings. Edited by G. A. Watson. X, 201 pages. 1976.

Vol. 507: M. C. Reed, Abstract Non-Linear Wave Equations. VI, 128 pages. 1976.

Vol. 508: E. Seneta, Regularly Varying Functions. V, 112 pages. 1976.

Vol. 509: D. E. Blair, Contact Manifolds in Riemannian Geometry. VI, 146 pages. 1976.

Vol. 510: V. Poènaru, Singularités C^∞ en Présence de Symétrie. V, 174 pages. 1976.

Vol. 511: Séminaire de Probabilités X. Proceedings 1974/75. Edité par P. A. Meyer. VI, 593 pages. 1976.

Vol. 512: Spaces of Analytic Functions, Kristiansand, Norway 1975. Proceedings. Edited by O. B. Bekken, B. K. Øksendal, and A. Stray. VIII, 204 pages. 1976.

Vol. 513: R. B. Warfield, Jr. Nilpotent Groups. VIII, 115 pages. 1976.

Vol. 514: Séminaire Bourbaki vol. 1974/75. Exposés 453 – 470. IV, 276 pages. 1976.

Vol. 515: Bäcklund Transformations. Nashville, Tennessee 1974. Proceedings. Edited by R. M. Miura. VIII, 295 pages. 1976.